+P93.5 .D38 1990

```
          Davies, Duncan.
P
93.5      The telling image
D38
1990
```

DATE DUE

The Telling Image

SCIENCE, TECHNOLOGY, AND SOCIETY SERIES

(Formerly Monographs on Science, Technology, and Society)

Editorial Board: THE LORD DAINTON, FRS (CHAIRMAN);
SIR GEOFFREY ALLEN, FRS; SIR WALTER BODMER, FRS;
SIR WILLIAM HENDERSON, FRS; PROFESSOR B. SUPPLE;
PROFESSOR S.A.V. SWANSON

1 Eric Ashby and Mary Anderson *The politics of clean air*
2 Edward Pochin *Nuclear radiation: risks and benefits*
3 L. Rotherham *Research and innovation: a record of the Wolfson Technological Projects Scheme 1968–1981; with a foreword and postscript by Lord Zuckerman*
4 John Sheail *Pesticides and nature conservation 1950–1975*
5 Duncan Davies, Diana Bathurst, and Robin Bathurst *The telling image: the changing balance between pictures and words in a technological age*
6 L. E. J. Roberts, P. S. Liss, and P. A. H. Saunders *Power generation and the environment*

The Telling Image
The Changing Balance between Pictures and Words in a Technological Age

Duncan Davies

Diana Bathurst

and

Robin Bathurst

CLARENDON PRESS OXFORD
1990

Oxford University Press, Walton Street, Oxford OX2 6DP
Oxford New York Toronto
Delhi Bombay Calcutta Madras Karachi
Petaling Jaya Singapore Hong Kong Tokyo
Nairobi Dar es Salaam Cape Town
Melbourne Auckland
and associated companies in
Berlin Ibadan

Oxford is a trade mark of Oxford University Press

Published in the United States
by Oxford University Press, New York

© Duncan Davies, Diana Bathurst, and Robin Bathurst 1990

All rights reserved. No part of this publication may be reproduced,
stored in a retrieval system, or transmitted, in any form or by any means,
electronic, mechanical, photocopying, recording, or otherwise, without
the prior permission of Oxford University Press

British Library Cataloguing in Publication Data
Davies, Duncan
The telling image: the changing balance between pictures
and words in a technological age.—(Science, technology
and society series, No. 5)
1. Man. Communication
I. Title.   II. Bathurst, Diana   III. Bathurst, Robin
IV. Series
302.3
ISBN 0-19-858339-7

Library of Congress Cataloging in Publication Data
Davies, Duncan.
The telling image: the changing balance between pictures and
words in a technological age/Duncan Davies, Diana Bathurst, and
Robin Bathurst.
(Science, technology, and society series: 5)
1. Visual communication.   2. Communication—Technological
innovations.   I. Bathurst, Diana.   II. Bathurst, Robin.
III. Title.   IV. Series: Science, technology, and the society series
(Oxford, England); 5.
P93.5.D38 1990   302.23—dc20
90-7598
CIP
ISBN 0-19-858339-7

Typeset by Cotswold Typesetting Ltd, Gloucester
Printed in Great Britain by
Courier International, Tiptree, Essex

Dedication

As Duncan did not live to complete and publish this book, his friends have done so as a tribute to him and in loving memory. The book is also a gift for Ann who has seen it through so many vicissitudes.

Foreword

The reflections and the speculations developed by Duncan Davies in this book evolved naturally from his day-to-day experience in industry. He joined ICI after graduating and gaining his doctorate at Oxford. In 1962 he was the founder and first Director of the Central Petrochemical and Polymer Laboratory in Runcorn. He was later Deputy Chairman, Mond Division, and then General Manager for ICI Research and Development. His interests in that period are shown by two books. The first, which he wrote with Callum McCarthy, entitled *Introduction to technological economics*, highlighted the desirability of technologists having at least a working knowledge of the basic economics of their industries. The second book, *The humane technologist*, in which he was joined by Thomas Banfield and Ray Sheahan, showed his perceptiveness, not only in understanding the impact of technology on the community, but also its impact on the technologist himself. He was, at that time, much concerned with the need for co-operation between industry and the universities, and devised the now familiar system of CASE Studentships. In 1977 he moved to the UK Department of Trade and Industry as Chief Scientist. After his retirement he spent four years as Chairman of British Ceramics, and was also President of the Society of Chemical Industry, and of the Research and Development Society.

Duncan was not only a fertile writer: he had powers of oratory too. He had the gift to paint pictures in speech. He could cross-examine and question one's understanding of a subject, and draw out in conversation an extra dimension. And he was always able to listen, and sometimes to redirect his own views. Yet in later life he began to feel that the human mind was somewhat constrained by the written or spoken word; that far more could be conveyed by pictures. He became aware that children can learn and develop very rapidly through visual experience long before they can read. Furthermore, pictures are not constrained by language.

Thus in his last years he became increasingly interested in the impact of information technology, in the way that information and ideas could be stored, manipulated, and represented in modern computing systems. Here too, pictures might be more powerful than text or tables.

As ever, concepts came readily to him, but getting the presentation and structure of this book into a satisfactory form required a great deal of thought, debate, and re-appraisal. He was just beginning to feel that he had the manuscript in reasonable order when he died.

Certainly in this period when he was wrestling with these ideas he had considerable impact on my colleagues and myself in ICI. We focused on the idea that information transmitted by text tended to be absorbed by the more academically inclined: operators on the other hand react better to pictorial presentations. Experiments are now in progress, for example, with fault-diagnosis and repair systems which incorporate video as well as tabular presentation of information for manufacturing and packaging lines. And this work will continue.

The text of this book has been completed by Duncan's friends. Though we cannot be sure how Duncan would have completed it, one thing is clear—he would expect each reader to use the text as a springboard for further excursions into the challenge of communication.

Geoffrey Allen

July 1988

Preface

Duncan Davies saw that great and influential changes are taking place in society, as the balance of the means of communication tilts progressively from words and numbers towards pictures or images. Feeling that the nature and import of these changes needed to be brought home to everyone, he was at pains to make this book intelligible to the layman with no special training in technology or science. The changes he saw—accelerating and increasingly computer-led—affect not just science and technology, but also commerce and industry, education, and the everyday world of the home, television, the supermarket, and the road user. Examples are the daily visual demonstration by the BBC of how weather evolves, international road signs, video-recording in surgery, computer-graphics in engineering, and the daily transmission of digitized pictorial data from satellites. Duncan felt that too few people are aware of either the scale or the speed of this shift in the balance toward pictorial communication, or of the problems posed by the need to store and retrieve pictorial data, or of the possible consequences for the future. Adapting to this transformation would, he believed, require profound modification of attitudes, and of the ways in which we prepare the young through education. Research and funding will also be needed; and here he applied his considerable experience in industry and government service to an assessment of desirable and likely trends in the coming years.

It was originally intended, of course, that Duncan should have sole responsibility for the text (his use of 'we' is figurative) and that Diana should have charge of the illustrations. This agreeable partnership ended sadly with Duncan's death in 1987. However, his family and the Oxford University Press subsequently expressed a desire that the book should be published, so Robin agreed to look after the text. Duncan had left only the briefest notes indicating his intention of elaborating, rearranging, and extending certain sections of the book; but with no details of how this was to be done. While every effort has been made to preserve his text, it was a first draft, and inevitably required some minor adjustment. Certain additions have been made to render some passages easier for the layman to follow. We are most grateful to Sir Edward Pochin for allowing us to quote extensively from his lucid exposition of the role of DNA and RNA in the storage and transfer of information. Since neither of us has

Duncan's broad and acute grasp of technological progress, we are specially grateful to Professor Jack Meadows for writing Appendix A on recent work in computer graphics. Duncan had intended to add to the section on the use of pictures in physics by giving a list of selected uses in an appendix, but he left no details to guide us. In the event, this omission is not as serious as might have been expected, because applications of pictorial methods in physics permeate much of the book. If the book seems biased towards chemistry or earth sciences, it should be remembered that Duncan was a chemist and Robin is a geologist. Finally, a word about the Bibliography. So far as we are aware Duncan had no plans for a list of further reading. Indeed his main reason for writing this book was his conviction that it would provide a text where none existed. He read widely in technical journals and reports, but was little concerned with the theoretical background, basing his text on a lifetime's experience in factory, laboratory, and managerial office. The few books he found really useful are referred to in the text, and are listed with a few additions in the Bibliography. From our own experiences we feel that readers are likely to find valuable further information in *New Scientist* and *Scientific American*.

The illustrations are intended to serve several purposes. They are grouped in an attempt to follow themes in the text, and, where possible, to give expression to historical development. It is hoped, also, that some of them will help the reader to an understanding of certain concepts that may be unfamiliar. The device of the long caption has allowed the insertion of explanatory and interesting information without interrupting Duncan's text. The range of possible pictures has necessarily been curtailed by the restraints of cost, time, and limited contacts (Duncan's support has been greatly missed), and reliance on donated material has been fundamental.

Lastly, we wish to thank Ann for placing her trust in us to complete this work of Duncan's.

D.W.B.
R.G.C.B.

July 1988

Acknowledgements

When I agreed to find the pictures for this book, it had not been written, and I little knew what would be involved. I am grateful for the help and encouragement given me by many people as I roamed through worlds of which I had previously been ignorant. The enthusiasm and interest of the many specialists who gave so much of their time to explain their subjects, and subsequently gave me pictures, not only impressed and educated me, but provided corroboration for Duncan's belief that we are further along the road to pictorial communication than is generally realized.

It is difficult to single out those who should receive my particular thanks, for the generosity has been so widespread. Miss Jacqueline Wilson initiated me into the arts of the picture researcher by giving me valuable advice on procedures and sources. Prof. G. J. Davies, Vice-Chancellor of Liverpool University, gave his blessing to a request that I plunder his institution for pictorial material, and Mr D. Bamber, the Public Relations Officer, and many members of the University have endured harassment as I have waylaid them and begged from them. I am grateful to Prof. D. J. Bacon, Dr M. A. Ball, Miss J. Carpenter, Dr P. H. Dangerfield and his colleague Dr D. Groves of the Liverpool Polytechnic, Dr P. Ditchfield, Dr J. Driscoll, Mr S Ferguson, Dr A. Henderson-Sellers, Mr R. Hunt, Dr R. H. Hunter, Dr J. Lucas, Mr W. E. Marsden, Dr J. L. Schonfelder, Dr W. R. Tyldesley, Mr J. Vaughan, Mr C. J. Veltkamp, and the staff of the Magnetic Resonance Centre and of the Central Photographic Service.

The Director of the B.B.C. Open University Production Unit, Mr P. Wilson, and his chief photographer, Mr D. Amy, gave me freedom to search their collection of pictures, and Miss H. Harrison of the Open University Library gave me the run of her domain.

A community of chemists formed a chain, and I was handed from link to link until I reached one who had the required illustration. Prof. J. Beynon, Prof. D. M. Blow, Dr B. W. Matthews, Dr S. Neidle, and Dr M. F. Perutz have been particularly generous in their efforts to assist me. Mr G. Chaloupka, Miss J. Davies, Dr N. Drake and Dr S. Mackin, Mrs D. Edwards, Dr R. M. Jacobi, Dr D. Marder, Mr C. Pearson, Prof. M. Rowan-Robinson, Dr A. Stride, Mr B. Tandy and Miss E. Amos, Dr J. Taylor and Dr S. Astley and Mr R. Waslewski have supplied me with interesting material which I have been delighted to use. I am grateful also to Mr A. Carter and to Mrs E. M. Jellis, Headteacher of Shorefields Community Comprehensive School, Dingle, Toxteth, Liverpool. Mr R. Fraser of Fraser Graphics drew Figs. 2.4, 2.7, 2.8, 2.10, and 7.4, and I am indebted to him for his suggestions and skill, and Mr C. Marsh of Recent Productions has been most helpful in directing me in the world of interactive video and computer symbols.

I have used some pictures which were amongst Duncan's slide collection, but for one of these, Fig. 7.11, I have been unable to trace the copyright. I think the transparency is one which Duncan collected from friends in industry. I sincerely hope that if the picture is recognized, the owners will accept with good spirit its inclusion in this tribute to a colleague and a friend.

Oxford University Press have steered me through many problems as they have arisen, and I have been most grateful for their expertise and friendly support.

I thank, also, those many friends and strangers whose photographs and diagrams I have been unable to use, particularly Mr R. Shingles, who willingly spent time devising mathematical diagrams.

Institutions and commercial enterprises have been generous in their replies to my requests, and museum staff unfailingly friendly. Amongst these I wish to thank the following: The Audio Visual Dept., University College, London: Fig. 8.3; The British Museum: Figs. 1.3, 2.9; Cambridge University Museum of Archaeology and Anthropology: Fig. 1.4; The Cancer Research Campaign at the Institute of Cancer Research: Plates 10, 16, 18; The Jet Propulsion Laboratory, Pasadena: Plate 12; The Laboratory of Molecular Biology, Medical Research Council, Cambridge: Plate 9; Liverpool City Libraries: Fig. 6.4; The National Remote Sensing Centre: Fig. 7.8.; The Natural Environment Research Council: Plate 24; National Radio Astronomy Observatory: Plate 25; The Satellite Receiving Station, University of Dundee: Figs. 4.4, 7.7; The Science Museum [London]: Figs. 1.2, 5.1, 5.3, 6.2, 6.5; Sheffield University and Health Authority, Dept. of Medical Physics and Clinical Engineering: Plate 15; Wolfson Image Analysis Unit, University of Manchester: Figs. 6.11, Plates 19, 29; Acorn Computers Ltd: Fig. 3.6; Apple Computer U.K. Ltd: Fig. 7.6; Alfred Marks Bureau Ltd: Fig. 3.7; Associated Press: Fig. 5.6; Central Electricity Generating Board: Fig. 5.5; Clyde Surveys Ltd: Plates 4, 21, 22, 23, Figs. 8.6, 8.8; Epic Industrial Communications Ltd: Fig. 8.2; General Electric: Fig. 6.10; Hulton Picture Company: Figs. 1.1, 2.5; Hunting Technical Services: Fig. 4.5, Plate 26; IBM. U.K. Ltd: Fig. 5.12; Innovations Mail Order Ltd: Plate 28; Interactive Information Systems Ltd: Fig. 8.1; Integraph (UK) Ltd: Plate 5; Kodak Ltd: Plate 6; Merseyside Passenger Transport Executive: Fig. 7.13; Nigel Press Associates Ltd: Fig. 7.8; Oxford Scientific Films Ltd; Philips Electronics; Plessey Controls Ltd: Fig. 2.12; SysScan U.K. Ltd: Plate 20; Telefocus, British Telecom: Figs. 7.2, 7.3, 8.10;

Acknowledgements

Unilever U.K. Central Resources Ltd: Figs. 2.13, 7.12; Videodem Interactive Systems Ltd: Fig. 8.5; Visual Data Systems Ltd: Fig. 8.4.

I also wish to thank the following for allowing me to include symbols associated with their organizations in Figs. 2.10 and 7.4; British Telecommunications plc (registered trade mark); The Campaign for Nuclear Disarmament; The International Olympics Committee; The Law Society; National Westminster Bank plc; The Open University; Penguin Books Ltd; Shell International Petroleum Co. Ltd; The Order of St John.

My thanks go to the authors and publishers for permission to reproduce material from the following publications:
J. B. Angevine and C. W. Cotman, 1981: *Principles of neuroanatomy*, Oxford University Press (Fig. 6.20c); F. B. Armstrong, 1983: *Biochemistry*, Oxford University Press (Fig. 6.16); R. G. C. Bathurst, 1971: *Carbonate sediments and their diagenesis*, Elsevier Scientific Publications (Fig. 5.2); R. Bellairs and E. G. Gray, 1974: *Essays on the nervous system*, Oxford University Press (Fig. 6.20b); C. G. Caro, T. J. Pedley, R. C. Schrotert, and W. A. Seed, 1978: *Mechanics of circulation*, Oxford University Press (Fig. 6.14); F. Close, M. Marten, and C. Sutton, 1987: *The particle explosion*, Oxford University Press (Figs. 5.8, Plate 14); J. Darius, 1984: *Beyond vision*, Oxford University Press (Figs. 5.7, Plate 11, 13); R. Dawkins, 1986: *The blind watchmaker*, Longman Group U.K. Ltd (Fig. 6.19); Dept. Education and Science, 1969: *Learning about space*, Her Majesty's Stationery Office. (Fig. 7.9); D. I. Gottleib, 1988: GABAergic neurons, *Scientific American* **Feb. 88** (Fig. 6.20a); I. R. Hatton, 1986: *Geometry of allochthonous chalk group members, Central Trough, North Sea* (Marine and Petroleum Geology, Vol. 3), Butterworth & Co. (Publishers) Ltd (Fig. 5.10); D. R. Hofstadter, 1980: *Gödel, Escher, Bach: an eternal golden braid*, Penguin Books Ltd. (Table 6.1); J. Holt, 1970: *How children learn*, Penguin Books Ltd; E. J. Lerner, 1984: Why can't a computer be more like a brain?, *High Technology* **Aug. 84**, Technology Publishing Co. (Fig. 6.20d); G. M. Loudon, 1988: Organic chemistry, Benjamin/Cummings Publishing Co. Inc. (Plate 16); A. Mauffret and L. Montardet, 1987: Rift tectonics on the passive continental margin off Galicia (Spain) (Marine and Petroleum Geology, Vol. 4.), Butterworth & Co (Publishers) Ltd. (Plate 27); *Oxford picture wordbook* Oxford University Press (Fig. 3.3); E. Pochin, 1983: Nuclear Radiation: risks and benefits, Oxford University Press; C. M. Powell, 1987: Inversion Tectonics in S. W. Dyfed, *Proceedings of the Geologists' Association*, **Vol. 98**. (Fig. 5.4); and C. P. Swanson, and P. L. Webster, 1985: *The cell*, Fifth edition, Prentice-Hall Inc. (Fig. 6.17b, c).

But it is the ongoing support of the day to day variety which is crucial to the successful outcome of an endeavour. Robin has written elsewhere that a husband's creations are rarely his alone. It gives me great pleasure to have an opportunity to reciprocate this compliment. I would have found it impossible to have completed this task without the balm of his patient support, his scientific knowledge, his skills as a draughtsman, and his willingness to take on the household chores while I have coped with the word-processor which he generously relinquished for the duration of my preoccupation.

June 1988

D.W.B.

Contents

1. The communications scene   1

    Scope and aims   1
    Practical aspects of the new options   3
    The five senses and their uses by different species   4
    The dominance of sight and sound in human communication and the place of pictures and words   5
    The basis for non-recorded communications   6
    Power patterns associated with recorded communication   6
    Attitudes and utility   11

2. The history of pictures, words, and numbers as records   13

    Background   13
    The limited impact of early pictorial recording   18
    Early codification of pictures into words and numbers   19
    The balance between different word-recording systems   23
    The Renaissance and the spread of words and numbers   26
    The impact of the digital computer   28
    The technological history of pictures   30
    The explosion of pictures   32
    Communication through television, video, and cheap colour photography   35

3. Pictures, words, and numbers in childhood   38

    Introduction   38
    The position before formal education   41
    The position in education up to fifty years ago   43
    Pictures and words in education now   46
    Pictures and words in future education   50

4. Pictures, words, and numbers: sorting and indexing   53

    Background   53
    Sorting words: further manipulation of the alphabet   56
    Sorting pictures: the absence of picture alphabets and syntax   57

The brain versus the computer: contrasting capabilities	61
Lines of attack on the sorting and indexing of pictures	64
The strengths and weaknesses of picture and word methods	67

5. Pictures and words in engineering, physics, and chemistry — 69

Background	69
Pictures in computer-aided engineering	71
Pictures in the physical sciences: chemistry	75
Pictures in the physical sciences: physics, astronomy, and earth sciences	78
Why video reports might be better than written ones	82

6. Pictures and words in biology and medicine — 86

Communication about and within biological systems	86
The role of artists and photography in the recording of natural phenomena	88
Surgical teaching and practice	92
Physic, pathology, and the recording of correct and incorrect function	95
Pictures in other species; directional perception; who can see what?	97
The impact of biochemistry and biophysics	98
Genetics and its cryptology and xerography	101
Brain science; artificial intelligence; molecular memories	107

7. Pictures and words in everyday life and study — 112

Background: the pervasiveness of television	112
How to display a picture	113
The return of pictograms and ideograms	115
Maps	119
Invisible pictures; infra-red; radio maps	122
Graphs, bar charts, and pie diagrams	124
Screen diagrams in control and guidance	125
Pictures in management, video-conferences, facsimile, time-lapse photography	127
Pictures in the delineation and indexing of inventions: patent search	128
Other pictorial symbols and clichés	129

8. The way ahead: the future of pictures — 132

Methods and approaches	132
Training, guidance, and education through simulation	133
Display of measurement and control	139
Decoration and entertainment, education, demonstration, illustration, explanation, display of the invisible	142

Contents xvii

Exploration and investigation, research, recording and storage of information	144
Communication and collaboration	149
Conclusions	151

Appendix A: Recent work on computer graphics 154
Jack Meadows

Appendix B: The role of computer graphics in the exploration of chaos 157
Robin and Diana Bathurst

Bibliography 160

Index 161

Plate 1 faces page 12. Plates 2 and 3 face page 45. Plates 4 to 30 appear between pages 76 and 77.

1. The communications scene

Scope and aims

The increased use nowadays of pictures and images, instead of words, in everyday life is a topic needing consideration by anybody and everybody, from teachers to businessmen, from research workers in science to technicians in industry, as well as by the man or woman in the street and in the supermarket. Human societies have been influenced profoundly by the different ways in which information can be conveyed or recorded, and much inefficiency, violence, deception, and unhappiness has been the result of misunderstanding or of inadequate information and communication. This book is about the changing balance between the use of words and numbers, on the one hand, and of pictures and images on the other. We shall pursue breadth rather than depth in the coverage and arguments, for we are all involved in these general issues: they are not the province of the specialist only. If we are to understand the changing uses of pictures, we need to appreciate their historical development. We are bound also to take note of the influence of education, especially on the young. We must examine the varied uses to which pictures are put in society, and we need to tackle the complex problems associated with their storage, sorting, and retrieval. These matters are considered in subsequent chapters. It will be useful for the moment to take a brief look at how the present situation has come into being.

We have communicated with one another by sight and sound for several millions of years; also by touch, smell, and taste. Other species have had this range of possible means of conveying information, but have evolved differently-balanced solutions. We have been able to record our experience, ideas, and messages for only about thirty thousand years. For the first twenty-five thousand of these we used only pictures. Their utility was rather limited because of the uncertainty of their meaning and the absence of conventions for guidance. About five thousand years ago, inventive minds allocated specific meanings to simple pictures that we now call pictograms. In these, the relation between the image and the object to which it referred was still discernible: stick men or flames of fire. Later there evolved ideograms, in which whole ideas and messages, or objects, were denoted by simple drawings. These were now too far developed for their origins to be visually identifiable. Then, about three thousand years ago, there arose amongst the Semitic people the even greater invention of the phonetic alphabet, in which ideograms were attached to the sounds that began the spoken indication of objects. From these there grew a cryptology, or organized encoding, so useful and powerful that part of the world from India through Western Asia and the Mediterranean to Europe (but not the Chinese or their neighbours) was able to record all of its knowledge and messages by the use of linear strings made up of about twenty or thirty basic elements (or letters).

Once this had happened the clerks (who could write) became very influential. However, about five hundred years ago, the development of printing in Europe destroyed the monopoly of the professional clerks, and reading and writing were opened up to whole populations. Pictures, as a serious means for the carrying of messages, receded into the background. More recently, the generation of powerful pictorial messages was brought within general reach by the advent of photography, TV cathode-ray-tube scanning, and computers. The pictures could be still or moving, two- or three-dimensional, monochrome or coloured, and representational or impressionistic. The introduction of these easily-generated pictures started just over a century ago, but gathered momentum only three decades ago.

Thus, there were 250 centuries when we had pictures alone and could not learn how to use them generally and effectively. Next there were 20 centuries when we learnt how to use formalized pictures (pictograms and ideograms) as message carriers. There followed some 15 centuries or so during which clerks made the great leap forward of alphabetic reading and writing. During the next 5 centuries, reading and writing thrust picture-communication into the background. However, over the past one-third of a century, the picture has suddenly and explosively become the main means for 'reading' and learning, but not yet the main means for the ordinary person's 'writing'. Not surprisingly, we are reeling from the shock. The return of the picture, although it has been the subject of intense political concern and enquiry, has not been thought about or studied at all broadly. We now have a rich array of options for recording experience, ideas, and messages. We are suffering from indigestion and *embarras de richesse*. The time has come for a little therapeutic self-discipline.

We have a curious tendency to regard the most open, transparent, and honest communication as self-evidently desirable. Pejorative words (obfuscation, deception, fraud) are associated with the interruption or corruption of messages. Yet, in the natural world, some of the most successful species have depended on the interruption of communication (for camouflage, Plate 1) or its falsification (to deceive others into acting in the interest of the falsifier), like the plants that persuade insects to carry their genetic material to the right place.

Alongside the spread of our ability to formulate general and qualitative messages, there has grown our ability to enumerate. The concept of number could only be very limited before its effective recording, on which mathematical manipulation depends. Precise numerical 'alphabets' began at the same time as pictograms and ideograms, but the powerful decimal notation of the Arabs did not arrive until after the verbal alphabet. The use of number as the exclusive power-base for various cliques (accountants, financiers, engineers) has persisted longer than the restrictive practices and power of the clerks, but this exclusivity is now fast disappearing, especially in Japan. The event in arithmetic that corresponds most closely to the impact of television and pictures in the non-numerical world is the spread of binary arithmetic as practised in the internals of the digital computer (on which the recording and printing of these words has depended). Words and numbers have merged together in the computer, as alphanumeric strings, for which we need a less clumsy word. There is still much difficulty attaching to the choice between conceptual thinking and argument, based on words, and quantitative thinking based on numbers. Yet this choice is a simpler one than that between words and numbers, on the one hand, and pictures on the other, which is the study we want to open up.

Practical aspects of the new options

The rapid spread of cheap and powerful microcomputers and mass-produced video-cameras and tape players, over the last three decades, has enhanced our ability to learn, and to a lesser extent to communicate, by producing and scrutinizing pictures of our ideas and activities. We are in the habit of recording and transmitting information as words, papers, and reports, sometimes illustrated with diagrams and photographs, but more often without them. We can now produce a video-tape as cheaply as a text report (if we cost honestly), and we will soon be able to include computer-generated animated cartoons almost as cheaply. There are new skills to be learnt, and the pleasures of using pictures can sometimes be greater than those of writing or using words. This is especially so if the events are not properly understood or are unfinished. When all is tidy, and there is a model that relates what has been done to the rest of human knowledge, then words, and numbers, and equations, may be more economical and powerful than pictures. This is so for Newton's laws, Maxwell's equations, or the more pungent poetry of Shakespeare. Even so, not everyone is happy with Occam's razor, and there are some who would far rather see *Hamlet* than read about the slings and arrows of outrageous fortune or the uses of a bare bodkin. There are also many who can understand motion and gravitation better from an animated graphics sequence, depicting planets and artificial satellites, than by setting and solving differential equations. For five hundred years, the victory of the word has inhibited the pictorially-minded. Now, if we wish, we can redress the balance.

Recent experience encourages this endeavour. First there is the success of television programmes on the natural world and a variety of other informative subjects—a matter quite distinct from the even greater success of television as an opiate. Secondly, educators (for example, in surgery) are making increasing use of tapes and closed-circuit television in formal teaching. Thirdly, five-minute video-tapes are proving extremely valuable in presenting the activities of laboratories to the politicians, non-technological senior officials, and high-ranking industrialists who at present do not know the stories, are unlikely to read reports, and lack the time (and sometimes the inclination) to pay visits. Such tapes, if well produced, can greatly improve relationships with paymasters. And fourthly, there was an occasion when a perceptive manager told those building a new plant at some distance not to bother sending reports, but just to send a set of photographs at the end of every week.

Our aim, then, is to examine the choice between the use of one-dimensional information (words and numbers or alphanumeric strings) and two-dimensional (and some three-dimensional) images, particularly in areas where image-formation is dependent on electronics or computers.

We already have machines and software that will assist us to start the job. We will find out more about their missing capabilities by trying out practical cases than by doing esoteric research (which will go on anyhow, partly, alas, for military purposes). The ordinary citizen is concerned and nervous about his computer illiteracy, and, quite rightly, sceptical, because of overselling and failures. If he has not himself had a gas bill for a million pounds, or nothing at all, then he knows someone who has. He has almost certainly suffered the frustration of trying to get an error corrected in a system containing too few people to carry out the necessary atypical operations. Although the way into personal computing is a smooth and easy one—if it is approached sensitively and sensibly—most attempts are badly guided and the manuals and teachers are not yet doing their jobs well. So the will to enter, and the conviction that it is worth while, are critically important. We want to try to add to the number of enthusiastic new potential picture-mongers.

The five senses and their uses by different species

Throughout the living world, individuals communicate through one or more of the five senses: touch, taste, smell, sound, sight. Among viruses and other micro-organisms, there is presumably no seeing or hearing, and tasting and smelling are merged into a rather general biochemical response to the presence of metabolites dispersed by their fellows. Characteristically, a culture of bacteria will not begin to multiply rapidly until critical concentrations are reached of certain bacterial products that are growth factors. Then, when growth in a highly nourishing medium has raised the population to a high level (often from transparency to opacity in a test tube), but before the food supply has been exhausted, it slows and stops, because of a response to poisonous by-products. Alongside these crudely perceptible signals and responses are many other more subtle messages, whose nature constitutes the subject-matter of microbiology. With this background, there can be no surprise that a rich organization of communications has built up in the plant and insect kingdoms through the use of metabolic products. Witchweed plants communicate through minute amounts of strigol, and other plants persuade insects to participate in their reproductive processes by play-acting, as in the plant that pretends to be a corpse so as to attract the flies. (This is an example of the essential and beneficial fraud and falsification already mentioned.) Indeed, the surprise is that the mammals, by the intensive use of eyes and ears, have made molecular communication between individuals decreasingly important as evolution has proceeded. Nor has physical touch gained ground, delectable though it may be in courtship. Could it be that we ourselves have increased our preference for sight and sound since the fourteenth century, when the five senses were given equal status in the tapestries depicting the Lady and the Unicorn, now in the Cluny Museum in Paris?

In passing, we may note that there is an interesting asymmetry between the different methods of communication which has had a considerable effect on their history. Some messages can be alternatively carried by two of the senses; others by one only. Some pungent molecules can be both smelt and tasted. Words, numbers, and music can be carried by sound or sight. With body language, there is a slight overlap between sight and touch. But pictures can only be carried by sight.

The dominance of sight and sound in human communication and the place of pictures and words

No more attention will be paid to smell, taste, and touch. Indeed, the size of our agenda makes it necessary to focus our gaze even more narrowly. Many visual and aural signals are expressed through dance, music, and cries of delight and horror. Nor is formal dance the main application of body-language, which we all use freely. It is curious that attempts have only recently been made to study the significance of our voluntary and involuntary movements. Some of our more imaginative policemen have been asking a choreographer (using ballet notation!) to define the kinds of movement that tempt a potential mugger to attack. Again, in music we know that our own varyingly tuneful and tuneless singing and whistling carry messages.

In this text the discussion will necessarily be confined to the records that begin with Cro-Magnon and the words inscribed on bone some four of five thousand years ago. If we use the conventional calendar-notation of one year to represent the period of terrestrial life, our study begins at one minute to twelve on December 31. The rich material of the rest of the year will be ignored, as well as much of the cacophony of communication in the action-packed final seconds.

The basis for non-recorded communications

Before there were drawn or written records, man's communications through sight and sound were based on observations of three kinds: visual images, visual and spoken primitive numerical concepts, and spoken words. All of these permitted recognition, conversation, and instruction through appearance, deliberate gesture, and partly deliberate facial expression. Such communications were the main methods for recognizing the individuality that allowed friends to be distinguished from foes, as David Attenborough has pointed out. But some gestures will have provided the dawning of measurement and number, for example by the holding up of fingers to indicate how many people were approaching or how many game animals were in the offing. Again, some conventional signs will have conveyed primitive verbal messages. Voice, however, will have been the principal channel for more refined uses of words and numbers.

Sight and sound communication thus overlap, in that both can represent ideas, words, and numbers, but they differ in the scope that they provide for precision. Without records, everyone has to develop his own system, based on rather crude general conventions (Fig. 1.1). The wheel has to be repeatedly re-invented. For obvious reasons, we know hardly anything about the processes of the early emergence of language or counting. Perhaps the best account (unavoidably speculative and imaginative) is to be found in the *Just so stories* of Rudyard Kipling.

Power patterns associated with recorded communication

More detailed history is the subject of the next chapter, but there are some general trends in the psychology and practice of communication that call for early comment. The recording of information must begin at the personal and

Power patterns associated with recorded communication 7

―――――――――――― **Idiosyncratic visual signals** ――――――――――――

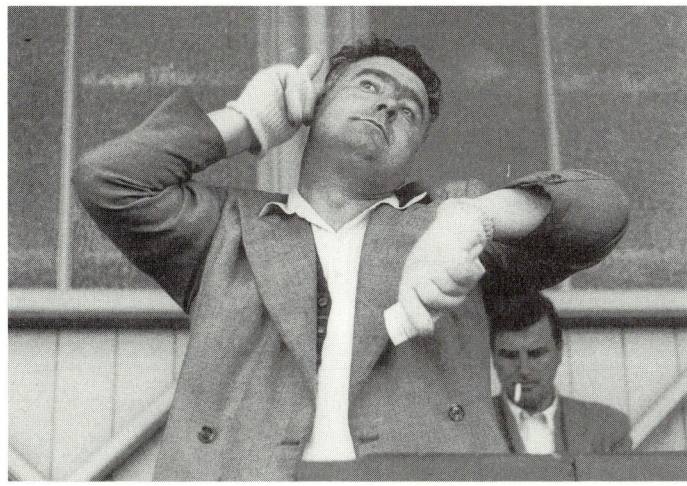

Fig. 1.1 Without records, everyone has to develop his own system. Esoteric system of tic-tac men. (Reproduced by courtesy of the Hulton Picture Company.)

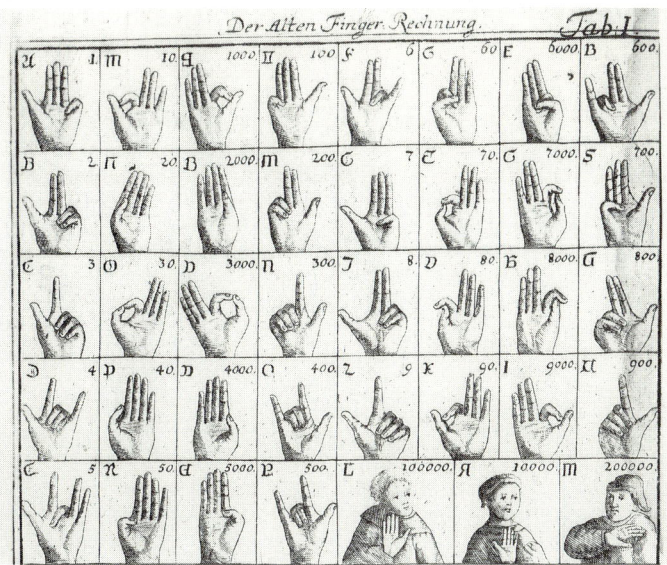

Fig. 1.2 Different communicators convey information in different ways. Bede's System for finger counting. (Reproduced by courtesy of the Trustees of the Science Museum [London].)

Primitive recorded messages

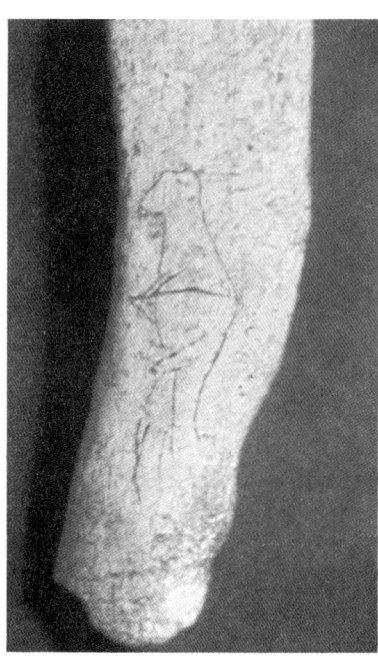

Fig. 1.3 Pictogram on bone from Pinhole Cave, Cresswell Crags, Derby/Notts border. About 14 000 years old. (Reproduced by courtesy of the Trustees of the British Museum.)

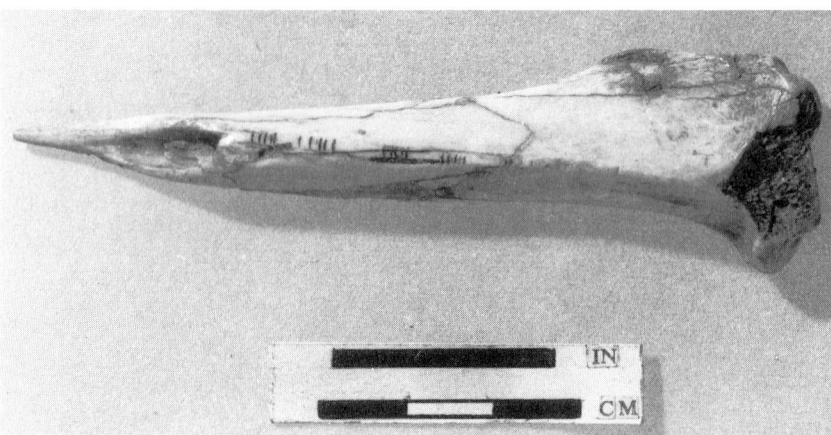

Fig. 1.4 Early recording of number. A notched bone from Gough's Cave, Cheddar, thought to show the days of the lunar month. About 13 000 years old. (Reproduced by courtesy of Cambridge University Museum of Archaeology and Anthropology.)

individual level. Someone wants to tell what he has seen or experienced, so he draws it (Figs. 1.3, 1.4). If, as a result, someone else is better prepared to cope with the newly-arrived mammoth, or herd of game, or herb that does good or harm, then the draughtsman will acquire reputation and influence. Different communicators will convey the same information in different ways (Fig. 1.2). Innovations lead to improvement in comprehensibility and transparency through standardization of images having similar meanings. This standardization of conventions is necessary if everyone is to use the pictures as the basis for a common method for recording. It is a pity that such standardization inevitably slows down innovation and further improvement (an experience common today in research and development, and mirrored in the incompatibility of our individually standardized computer software).

We do not know how long it took for the immediately comprehensible image to become formal. Once this had happened, meaning no longer had to be learnt again each time. Did the development of standards and comprehensibility in written language require parallel developments in spoken language? We do, however, know that the use of linear words (and numbers) has been a specialized skill, available to no more than a minority, until very recently. The majority of people learnt formal and standardized spoken language, but for comprehensible recording they depended on pictures and sculptures. All had voices and ears, but all could not learn to write or read because of the absence of the time needed for the copying of books. The European church used words and numbers for internal communication among the monks and higher clergy, but employed images for telling the people. To what extent were Brueghel's *Tower of Babel* and *Triumph of Death* explicitly designed for propaganda purposes?

The possible arrival of printing must have appeared to the intelligent cleric as a dangerous potential threat to ecclesiastical power. Again, we do not know whether this invention was held back when, in the twelfth and thirteenth centuries, the church was at the zenith of its authority. Certainly, the actual arrival of the press coincided with a decline in the power of the monasteries. Printing has continued to threaten authority up to the present day, as government seeks to impose injunctions on newspapers restraining them from publishing sensitive information (Fig. 1.5).

The more difficult ideographic scripts of eastern Asia may have been more effective than alphabetic writing as a means for preserving the power of the bureaucrats. In China, popular education has taken longer to develop. In Japan, economic need and opportunity for general technical education has helped the introduction of European languages or European-style alphabets as part of the solution. Yet again, as we shall see, the whole pattern of development of Indo-European writing might almost have been designed for the arrival of the digital computer with switch-based memory. For European language turns two- and three-dimensional pictorial perception, and derived

Perceived threat of the press and universal literacy

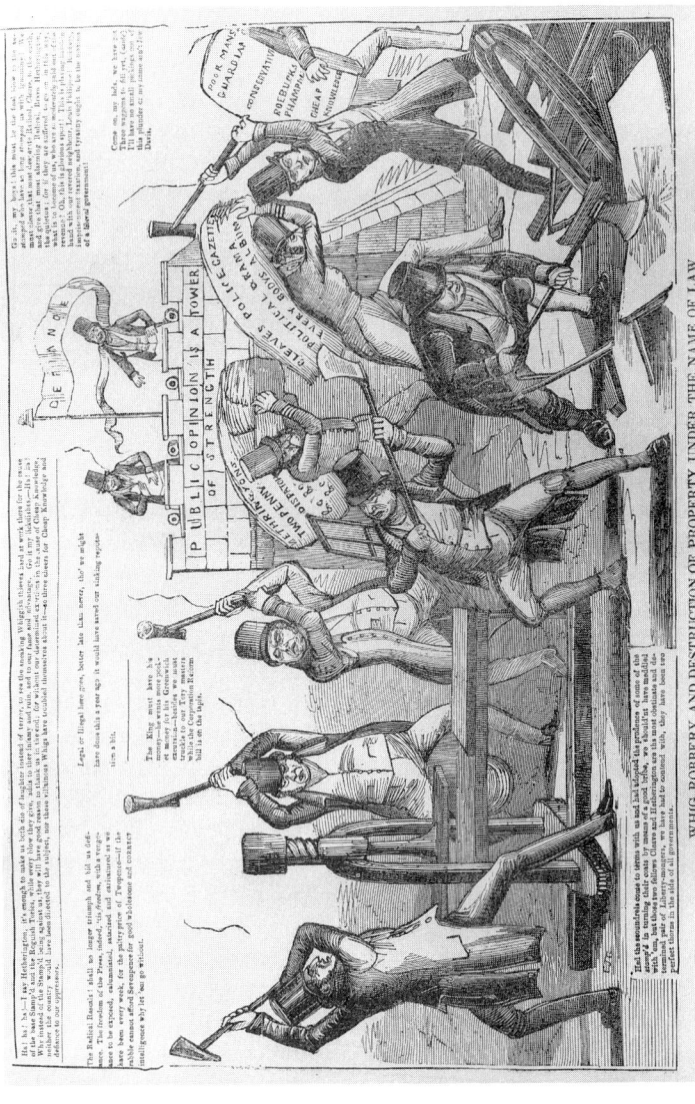

Fig. 1.5 1835 cartoon of the seizure and destruction of Hetherington's Press. (Reproduced by courtesy of the BBC Open University Production Centre.)

abstraction, into one-dimensional script, which is exactly what the present computer needs, both for its operation and for the organization and indexing of its material. Pictures can, of course, be handled by computers, but their codification, indexing, and retrieval (which goes under the name of 'pattern recognition') are still very primitive. (The molecular memory of the brain does a much more sophisticated job.) So linear strings of information at present occupy a doubly-strong position in our culture—because of the ease of printing, and because we invented computers to deal with numbers by translating them into binary code (Chapter 4), a process which could most readily be extended to words in the linear Indo-European language. By contrast, the raster scan that gives rise to the image in a cathode-ray-tube (whether pictures, words, or numbers) does not depend on linear syntax but on building up the appropriate pixels, or a mosaic of spots (Chapter 4), to make symbols or a picture through a process of systematic raking or sweeping of a succession of parallel lines, one on the other, until the whole screen is covered.

Attitudes and utility

The market, in theory, links the capabilities provided by the computer makers and software writers with the needs of the common customer. But the mechanism does not work very well when the customer lacks the knowledge and confidence to state his needs. As yet, people do not know, or recognize, the use that they could make of a capability to generate and handle good pictures. They do not yet perceive the needs or opportunities that they might articulate or formulate. We may ask what has been achieved by simply making computers and video-cameras as readily available as Japanese colour cameras? At present, children have little computers for trial at school and to play with at home. If inquisitive, they can try their hands at programming. Secretarial staff are using computers in the office, though often only as labour-saving typewriters. Ordinary accountants, book-keepers, and engineers are using systems that are cheap for business but dear for the individual. Ordinary teachers are beginning to help with all these processes. But there is very little sign of the 'information revolution', 'the paperless office', '*la télématique*' or 'the high-technology cottage industry' about which so much has been imaginatively written.

Where should priorities be applied? Are we held up by the input-output system? Is the QWERTY keyboard something that we shall all get used to, as children become accustomed to it as the means of playing games? Are we articulate enough to want voice-entry or is this something, like the dictating machine, that will only be used by the very self-confident? Would prompt-lists and touch-pens help? Or will confidence develop faster if simple computers become part of ordinary telephone terminals, for leaving messages or inviting twenty people to a party? If so, then we must do more about low-cost standardization.

Will picture-drawing become one of the important ways into general computer-confidence, or will people normally learn computer technique by starting with the handling of texts and numbers? Will cheap and simple big-pixel techniques be attractive, or will ordinary people only feel at ease with something more sophisticated, and, at present, expensive? How many people are comfortable with Teletext or Prestel? Will the active making of video-tapes encourage the use of television for explicit and formal learning (as opposed to passive watching) and become a general skill? Or will it remain specialized? Will the video-system (tape or disc) be connected with the computer closely or loosely? Will a general appreciation of the power of the picture evoke a new interest in drawing with a pencil, and will this be the main response of the ordinary man? Will this interest be helped by more formal ideography and iconography, such as that of the traffic sign? Above all, can the use of pictures be warm rather than cold, and friendly rather than threatening? Will it all be unwelcome to the person who wants to spend a lot of time with his own thoughts? Are we wrong in assuming that an increase in the ability to communicate will be beneficial, or will it merely make life uncomfortable? Will picture communication introduce yet another kind of social division, by further disabling those with slower or more limited brain-power?

It will not be useful to answer these questions with nothing more than philosophy. At the risk of irritating the specialists for whom theory is important, the aim will be to make practical suggestions about the steps that will generate practical interest. These should help and encourage numbers of people to become involved and active in those steps that will be useful for ordinary and everyday purposes.

The interruption of visual communication

Plate 1. A camouflaged moth. (Reproduced by courtesy of the BBC Open University Production Centre.) See p. 2.

2. The history of pictures, words, and numbers as records

Background

David Attenborough, in his book, *Life on Earth*, calls men the 'compulsive communicators'. He says, 'Once we thought that we were the only creatures to make and use tools. We now know that this is not so: chimpanzees do so and so do finches in the Galapagos that cut and trim long thorns to use as pins for extracting grubs from holes in wood. Even our complex spoken language seems less special the more we learn about the communications used by chimpanzees and dolphins. But we are the only creatures to have painted representational pictures and it is this talent which led to developments which ultimately transformed the life of mankind'. And *Life on Earth* is itself a splendid illustration of the balance that we are studying. In the form of a television documentary of a scientific analysis, it was compelling in a way that no book could ever have been. Yet the text is also available as an attractively illustrated, traditional book, which scores over the television version by being easier to browse in and by enabling the reader to go over arguments many times in order to test and consolidate new knowledge. Television and books are compatible and synergistic.

Although man may be the first conscious creator of records (whether informal pictures or formal verbal or numerical symbols) he is a very raw apprentice by comparison with the extreme precision involved in biological reproduction. One of the main purposes of writing or drawing is to encode knowledge. This makes it possible for a job to be done quickly by calling on all relevant experience. Otherwise the procedure has to be invented afresh every time and the same experimental failures may be repeated endlessly. Long before man wrote and drew, the processes of biological reproduction needed a capability for faithful replication: the genetic record involved in the replication of DNA (Chapter 6) is incredibly accomplished in this respect. The information in the first cell from which a multicellular organism grows permits the reproduction of the species without mistakes or hesitation. The biological reading of this record is implemented through communication using chemical signals. This write-and-read process has taken over a billion years to evolve. The biological library now has the recipes for the making and operation of single-cell organisms, plants, insects, reptiles, and mammals. Included in the instruction manuals are the data for growing and reproducing the organs for communication through sight and sound, with whose use we are concerned. We will return to the balance between pictures and words in the elucidation and imitation of biological recording and communication in Chapter 6.

Man has been developing his recording skills for no more than thirty thousand years or so—or about one-hundred-thousandth of the time that has been available for the evolution of the genetic code. Considering this, his performance has not been bad. For the first 250 centuries, a few individuals were painting visual images and symbols that codified thoughts of many kinds. During this long dawn of recorded communication, there was much experiment with symbolic drawing, some connected with religion and some with fecundity and reproduction. These trials have continued right up to the present and take forms varying from the large drawings in gravel at Nazca in Peru to the abstract art of Mondrian and Pollock (Figs. 2.1, 2.2, 2.3). But we do not properly understand the purpose of the early drawing. Only for a little more than five thousand years has man used numbers and words to provide disciplined records of his gestures, thought, and talk. Writing underpinned the commenting and counting that he found useful. He thus acquired a capability which helped with commerce and the management of the community and its flocks of animals. Although the study of the whole subject of communication—unconscious and deliberate—must contain many clues for the study of the records relevant to ideas and messages, it is too big a subject for inclusion here. The history of art alone embraces scores of academically distinct branches of scholarship.

Fig. 2.1 Scandinavian rock carving. Bronze Age at Vitlycke, Sweden.

Fig. 2.2 Lines in the Nazca Desert, Peru. The bird's bill reaches a set of lines, one of which points to the rising sun on December 21. (Reproduced by courtesy of Roman Wasilewski.)

Fig. 2.3 Painting by Jackson Pollock. Twentieth century. (Reproduced by courtesy of the BBC Open University Production Centre.)

Background

Symbolic drawing

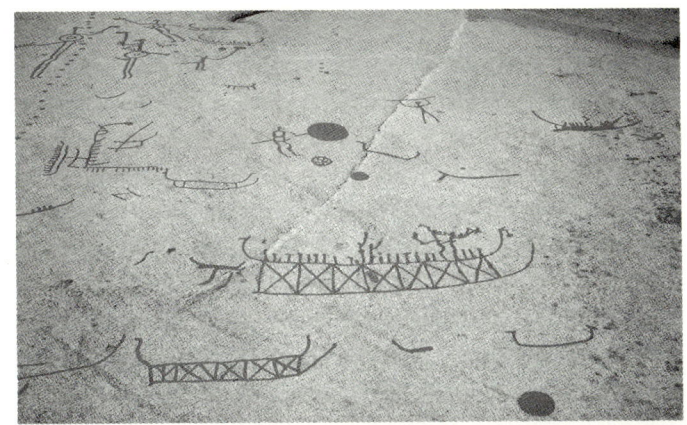

Fig. 2.1

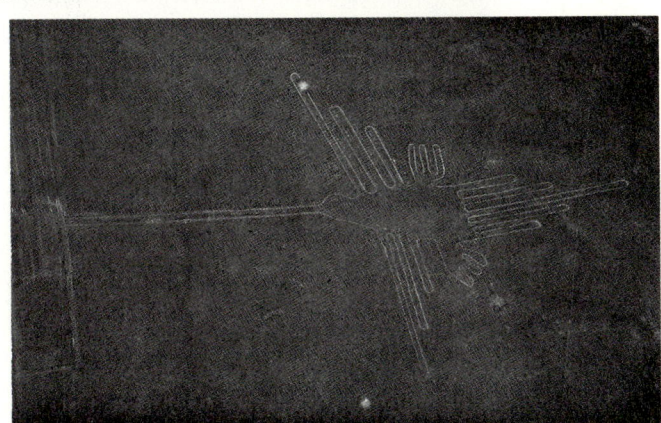

Fig. 2.2

Fig. 2.3

16   *The history of pictures, words, and numbers as records*

──────────────── **Skilled pictorial message** ────────────────

Fig. 2.4 An early 'thinks' cartoon. The rich man and the poor man contemplating Christ on the Cross. Keldby Kirke, Denmark. Fifteenth century. (Reproduced by courtesy of Fraser Graphics.)

Background 17

———— **Skilled pictorial message** ————

Fig. 2.5 Fougasse poster for the anti-gossip campaign 1940. (Reproduced by courtesy of the Hulton Picture Company.)

The limited impact of early pictorial recording

There would seem to be two reasons why pictorial recording, for its first 250 centuries, contributed so little to communication. First, as mentioned in Chapter 1, it did not offer the standardization of representation that was so badly needed for codification. Speech must have done far more, both for words and numbers. Secondly, painting required various rare talents. These included skills in drawing or colouring, and expertize in the making of pigments and implements. Only a small minority, in a hunting and gathering society, seems to have been able to acquire the necessary competence. Even now, it is only a minority who can draw or paint convincingly, even though the materials can all be bought cheaply at the corner shop. Most importantly, picture standardization led, in due course, to formal pictograms and to the generally useful records that, over the last three millennia, have swept past pictorial methods, but not obliterated them. Pictures have remained as means whereby a skilled minority has been able to convey powerful messages to the general public (Figs. 2.4, 2.5). It has been helped by printing, which has mobilized a clarification and fortification of messages in books and periodical publications.

────────── Picture message with limitations ──────────

Fig. 2.6 A painting of The Massacre of the Innocents showing no emotion. Elmelunde Kirke, Denmark. Fifteenth century.

Early primitive painting did not record gesture, emotion, or identity through facial expression or movement (Fig. 2.6). In this kind of communication it failed. A further difficulty with the early use of pictures must have been that some practitioners will have produced results that were actually harmful. An unskilfully produced picture may fail to convey a message because of unattractiveness (which causes the viewer to ignore it), or because of obscurity (which prevents the viewer from understanding it, or which may even create the wrong message).

Early codification of pictures into words and numbers

Before real progress in written language could be made, a conventional codification was needed, that is, simple symbols of a learnable kind. These were provided by the Indo-European phonetic alphabets and the Arabic numerals, which were particularly readily used and taught. Initially, the learning might have been confined to a limited clerical class. Later, it spread throughout society as education became a social need.

The earliest codification of drawing that we can now recognize is that for the representation of numbers. This seems to have emerged, after about 200 centuries of pictures, around 4000 BC, and to have been quickly found useful. As time went on, different numerical scripts emerged. Before these underwent natural selection, a preliminary local standardization must have substantially widened the range of users. This very interesting subject has attracted relatively little study and attention, but one of the most sensible outline treatments is in a popular book, Lancelot Hogben's *Mathematics for the million* (1936).

The ancient Egyptians and Babylonians had practical astronomical and engineering concepts for the measurement of time and dimensions, and the ancient Greeks pressed ahead with philosophy carried into pictures in the form of geometry. In later history the Romans, whose administration would have benefited from better accountancy, were held back by a clumsy numerical notation. Elsewhere, Mexicans devised building systems more advanced than the Egyptian, while the short-lived and powerful Inca empire (with no written script at all) made do with notation based on knotted string. The Chinese had relatively little mathematics and yet became good engineers. Fig. 2.7 shows some examples of the symbols and systems used to record number during these early years. They were all clumsy in one way or another. It was the Arabic system of numerals, with the Hindu introduction of the symbol 'zero', that provided the key. These developments, together with the intellectual experimentation of the Renaissance and the dissemination of ideas through printing, underpinned mensuration and gave rise to the study of measurement

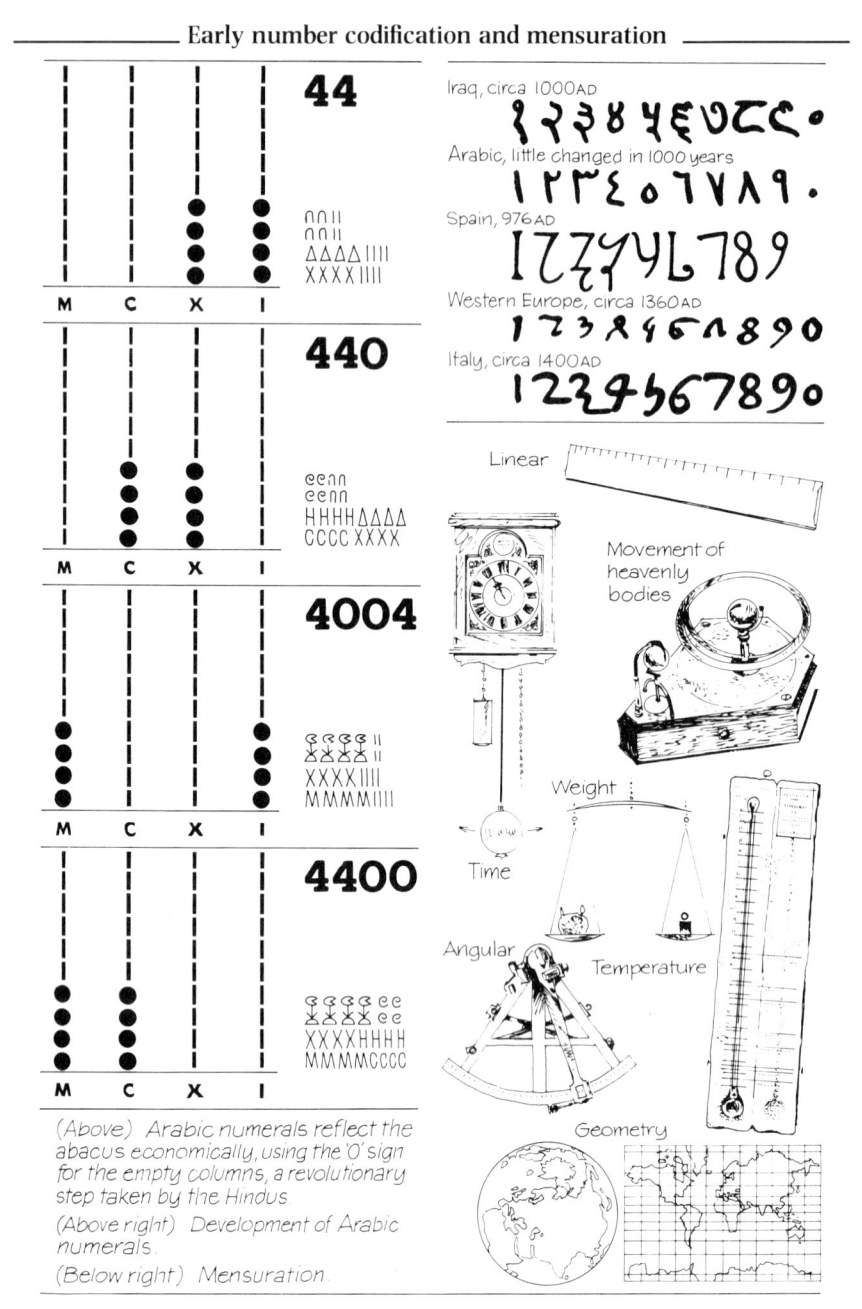

20   *The history of pictures, words, and numbers as records*

Early number codification and mensuration

44

Iraq, circa 1000AD

Arabic, little changed in 1000 years

Spain, 976AD

Western Europe, circa 1360AD

Italy, circa 1400AD

440

4004

4400

Linear

Movement of heavenly bodies

Weight

Time

Angular

Temperature

Geometry

(Above) Arabic numerals reflect the abacus economically, using the '0' sign for the empty columns, a revolutionary step taken by the Hindus

(Above right) Development of Arabic numerals.

(Below right) Mensuration.

Fig. 2.7 Numerical scripts and concepts for measurement. (Reproduced by courtesy of Fraser Graphics after L. Hogben *Mathematics in the making.*)

correlation, and thence to algebra and calculus. By the eighteenth century, the sophistication of the manipulation of number paralleled that of the written word. After this, general education caused both to be popularized and used widely.

Alongside codified numbers, there emerged incipiently codified pictures, to represent non-numerical ideas, items, and events. The pictograms of the fourth millennium BC have been found in quantity by archaeologists, but their decoding can only be speculative. There obviously can be no Rosetta Stone of the period, on which whole pages of pictograms lie alongside the same messages in Greek! Figure 2.8 shows examples of this form of representation and of the ideograms—or even more formalized pictures—into which they evolved. It has been possible to decode some of these, because of the overlap in time between the use of ideograms and the early alphabetic scripts. This overlap permitted decoding by backward extrapolation of our present linguistics. After twenty more centuries, at about 1000 BC, there came the highly innovative step whereby a small range of about thirty particular ideograms became accepted to represent the initial sound of the word or group of words expressed. This writing (whether of words or numbers) was mainly used to encapsulate messages that had previously been conveyed only by voice as speech or sung in verse. Even if badly written, or mis-spelt, it was much more easily understood than the contemporary, approximate, informal and non-codified pictures.

In the Indo-European areas (but not in China and its environs) the phonetic alphabet came into being. Its simplicity and ease of learning caused it to be accepted rapidly by the scribes. The Hebrew and Greek alphabets quickly secured adoption: the Roman and the Cyrillic scripts were born out of Greek. The study of the alphabets, and of the ideograms that can be reached through them, can take us back to 1500–2000 BC, but gathers momentum only from about 1000 BC. Scholars such as I. J. Gelb and David Diringer (who founded the Museum of the Alphabet in Cambridge) have written in depth of the later developments, and have done much to help our understanding of the earlier ones in which our present interest lies.

There was rapid recognition of words and numbers as a power base. This new power was used to considerable effect by various classes of writers and clerics. Within two or three millennia, in the civilizations of Babylon, Egypt, Greece, Rome, and Medieval Europe, some were using the ability to write books, poems, or plays to influence substantial audiences (there were not many readers). Others used the same skills to strengthen their political or military positions, or to manage affairs by communicating among themselves as part of the conduct of their administrative or priestly functions. Sometimes, the rulers and the priesthood acted as patrons of the painters, sculptors, and stonemasons, who were called upon to depict ecclesiastical and moral subjects in such a way as to help convey the stories of the scriptures or important

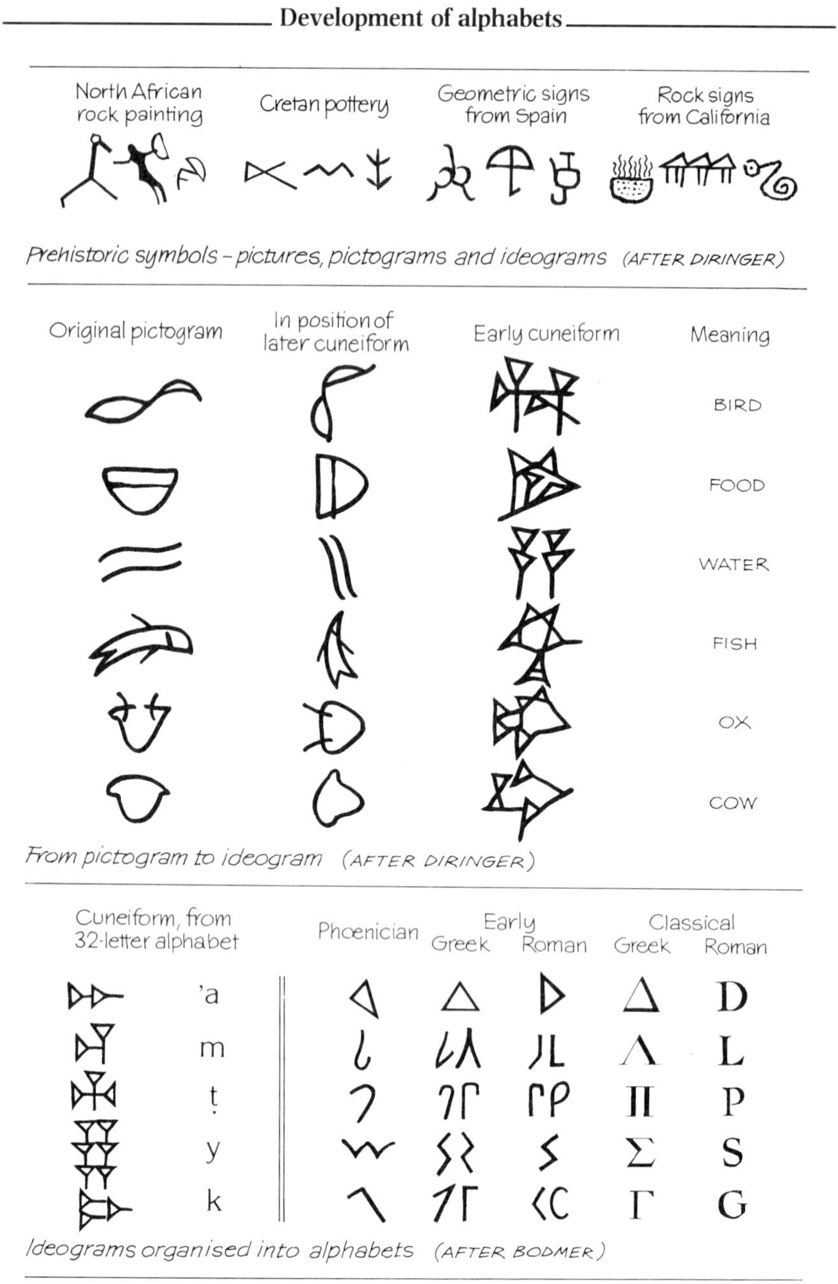

Fig. 2.8 From pictograms to letters. (Reproduced by courtesy of Fraser Graphics after D. Diringer *Writing* and F. Bodmer *The loom of language*.)

messages about the requirements of faith and morals. These craftsmen were employed as the servants of the clerics or of the rulers of the state (from whose authority the clerics were often exempt: they could sometimes unseat rulers who did not suit their purposes). Words and pictures were used in a balance which helped in the preservation of the status quo, but it was the wordsmiths who ruled the roost.

The balance between different word-recording systems

There are, of course, many possible methods for creating and using alphabets and symbols to record and transmit speech and ideas (Figs. 2.9, 2.10). One such is the syllabary, exemplified in our own time by the totally phonetic script of shorthand notation, which simply writes speech as it is heard, incorporating all the nuances of dialect, and accordingly displaying differences between the same sentence spoken by people from Scotland, Brooklyn N.Y., Ireland, Jamaica, Yorkshire, Papua, and Mayfair. Near to this is the etymological script of the Indo-European languages, which is strongly based on sound. This is now standardized and systematized in ways that help relate words and sentences to one another and to their origins. The representation in writing of a given sentence is the same everywhere, despite quite substantial differences in sound. (Words, of course, may vary synonymously: one can say 'cross-cut saw' or 'pull-him-he-come-push-him-he-go-brother-belong-axe', according to choice.) There are several systems for the use of the blind that can be felt with the fingers. In contrast, there are ideographic scripts and international ideograms. The former include the thirty-three thousand characters of Chinese. The latter embrace the growing array of technical and international ideograms, carrying messages about roads, traffic, and which sex may use a particular door.

Frederick Bodmer's book *The loom of language* (1944) contains a most attractive hypothesis about the reasons underlying the differences between the economy of the Indo-European near-phonetic alphabets of about twenty-five letters and the rich and much more varied Chinese and Chinese-related symbols. Chinese syllables, he says, are mainly open-ended (that is to say, consonant-sound plus vowel-sound, for example, 'me', 'to', 'by'). The total number of such monosyllables is only about four hundred, insufficient to form a language from monosyllabic words. The solution that evolved was the multiplication of words, both by increasing their number and at the same time applying to them different inflections, with much 'homophonic' overlap. Words that sounded basically the same differ by being said or pitched differently. The system of writing then had to distinguish between these homophones, and did so contextually. By way of illustration, Bodmer cites the

24   *The history of pictures, words, and numbers as records*

──────────────── **Other encoding systems** ────────────────

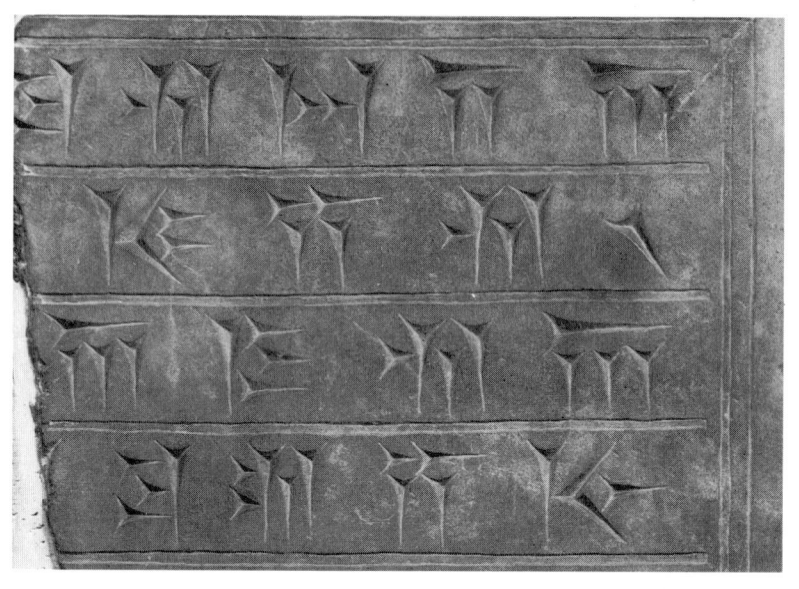

Fig. 2.9 Part of an inscription in Persian cuneiform script. 485–465 BC. (Reproduced by courtesy of the Trustees of the British Museum.)

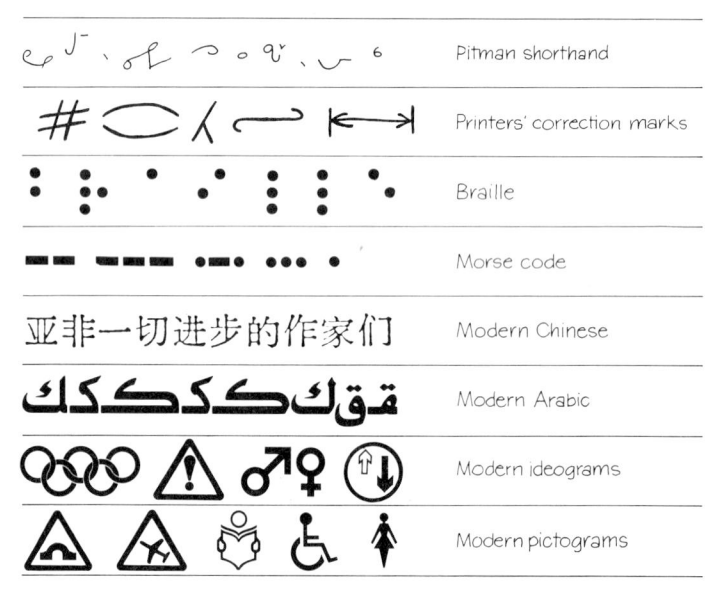

(shorthand symbols)	Pitman shorthand
(correction marks)	Printers' correction marks
(braille dots)	Braille
(morse dots and dashes)	Morse code
亚非一切进步的作家们	Modern Chinese
ككككقك	Modern Arabic
(olympic rings, warning, gender, arrows)	Modern ideograms
(pictogram symbols)	Modern pictograms

Fig. 2.10 Systems is use today. (Reproduced by courtesy of Fraser Graphics.)

words 'boy' and 'buoy', which in Chinese would have two composite ideograms, each containing the same homophone base, but in addition a male symbol for 'boy' and a marine symbol for 'buoy'. Hence all the ideograms.

The good fortune of the Indo-Europeans was to have syllables that were often closed-ended (that is, one or more consonant-sounds plus vowel-sound plus one or more consonant-sounds, for example, 'pet', 'tramp', 'most'). This meant that a purely phonetic alphabet having twenty consonants and five vowels (and with no contextual element) was quite adequate for a rich language. It could represent many thousands of monosyllables, and, with some additional aggregation of these to polysyllabic words, it gave a script with relatively few confusing overlaps of homophones. Such a script is a one-dimensional string, whereas the Chinese character is two-dimensional: one dimension being sound and the other being the contextual meaning.

Our linear and phonetic alphabet has given us two major benefits. First, it is easy to learn and suitable for building a society in which most people can read and write. Secondly, it has made it easy to index material, and, thereby, to prepare for the far-reaching benefits that flow from the computer-handling of such indexes.

Early indexing and lexicography were greatly helped by a syntactical accident. In nearly all the Indo-European languages, the changes in words consequent on the conjugation of verbs or the declension of nouns and adjectives are made at the end of the word, rather than the beginning. This means that the order of entries in the index and dictionary can best be based on the beginning of the word, starting with the words beginning with aaa, then going on to aab–aaz, and then to aba–abz, and so on. The tense or mood of the verb or the case of the noun or adjective does not significantly affect the position of the word, so that related adverbs, adjectives, verbs, and nouns occur near together in the index and dictionary. The importance of this stroke of luck can be gauged by looking at the Celtic languages, where letters at the beginnings of words can inflect grammatically. In Welsh the word for 'father' can be *dad, tad,* or *nhad,* with consequent difficulty in its location in the dictionary by a learner or non-speaker of the language. A computer could be programmed to cope with rules of this kind, but the whole process of language-sorting over the years, including computerization, would have been much more chaotic if grammar sometimes altered the beginning of a word, sometimes the middle, and sometimes the end.

There are, of course, scanning methods for turning images and pictures into the one-dimensional data strings that can be handled by a silicon-switch computer, as we shall see later in the chapter; but they are more complex and do not lend themselves to simple indexing.

The development of the Indo-European languages has also prepared us admirably for the use of the digital computer based on memories made up from two-position silicon switches. This operates by translating all words and decimal numbers into binary notation (Fig. 2.11, p. 29), after which they can be processed and indexed by electronic circuitry. We shall discuss computers in more detail in Chapter 4.

The Japanese, on the other hand, were unlucky. They used polysyllabic words (*Yo-ko-ha-ma, Na-ga-sa-ki*), which, although composed mainly of open-ended monosyllables, provided enough variety for a phonetic script, easily learnt and usable and indexable by computers. However, because they lived in the cultural shadow of the Chinese, they developed an ideographic script. Only recently have they turned to a new script which is phonetic, linear, and suitable for typewriters and computers.

The Renaissance and the spread of words and numbers

Returning to the historical development of the balance between words and pictures in Europe, it seems that the convenience of our script, with its phonetic base, was mobilized by printing and liberated from the monasteries by the Renaissance. As a result, near-universal literacy was, within only five centuries, achieved by 1950 throughout Europe and in other territories, including North America, which had acquired European culture and languages through colonization. The process was far from straightforward, but is well worth while tracing because of its importance in providing the reasons for the familiar dominance of words and numbers over pictures up to about 1950.

First, liberation from ecclesiastical control of learning was not the only result of the Renaissance. Another was diversity, which heralded obstacles to communication. Latin ceased to be the predominant language of the scholar, with the immediate growth of new local pride in local languages such as English, French, German, Italian, and Spanish, and the formation of a *Europe des patries*. Each of the several European countries followed a different pathway to near-universal literacy, and the balance between literacy and numeracy varied from region to region. However, although communication across frontiers may have been reduced, communication within countries was greatly stimulated. National consciousness led to the flowering of national capability on varying time patterns.

In the field of literature and words, English made a quick start, with the vigour and popularity of Shakespeare and the many other Elizabethan playwrights, poets, and men of letters. Standardization was helped, from 1500 onward, by the founding of the Grammar Schools. The major incentive to educate most of the people came with the Industrial Revolution from 1840 onwards. A more urgent approach to literacy in English developed in America, where universal education was made especially relevant by the shortage of trained people in a large and under-developed continent. Germany was held back by the Thirty Years War, and Spain by economic collapse brought on partly by the burden of the ill-starred attempt to retain control of the Netherlands. France enjoyed a golden era of literature and drama in the seventeenth century, followed by a very systematic process of dissemination of literacy as one of the consequences of the Revolution and of the policies of Napoleon.

In the field of numeracy there was more uniformity, perhaps because of the existence of the common language of the Arabic numerals and the international syntax of mathematics. Pure mathematics advanced in the seventeenth century in Britain, France, and Germany. Applied mathematics, built on foundations laid by Newton, was of most benefit in Germany (and to some extent in France), where it helped to create advanced engineering. This development was necessary to continental European countries where the economies grew differently from that of Britain with its availability of cheap raw materials. Manufacture, therefore, needed to be cheaper, so as to make the final products competitive. This trend, in turn, tended to encourage the growth of mathematical abilities in the whole population. Nevertheless, leaving aside a certain amount of lagging in countries like Spain, and international variation in the balance between words and numbers, overall progress in the numerical communications of the ordinary man was spectacular.

If an international script has done all this for mathematical capability, where there was virtually no oral tradition, have letters and written words done as much for verbal communication? Despite the beauty and volume of literature, they probably have not. In the first place, the word-based oral tradition was bound by much greater constraints. In the second place, word-language, unlike mathematical language, has not yet become international. Nor is the philosophy and logic of word-strings as unified as that of number-strings. Moreover, the mathematical manipulation of numbers follows precise rules, having no ambiguity. The manipulation of words, on the other hand, can leave much scope for interpretation. As long as there was an authoritarian church that did not tolerate heresy, then words and philosophy remained reasonably integrated, because ambiguity in meaning could be kept to a low level. With freer thinking, philosophy had to deal with different premises, and evolved different moral and ethical conclusions. The meaning of words assumed greater importance, and philosophers since Wittgenstein have retreated into this field.

It is, however, the string of words that has the most nicely balanced options as to the means of transmission: speaking and writing are almost equally powerful, although the precise assembly of words for a good lecture is appreciably different from that for conveying the same information on paper. The reading of a text that is an excellent written presentation is usually a recipe for a stilted and dull lecture, and the tape-recording of an engaging and witty conversation can make poor reading. By contrast, in the handling of numbers, writing is more powerful than talking, though the spoken word is much more powerful than the spoken number. Talking about numbers is useful, of course, especially in teaching; but it is usually necessary to have a 'visual aid' in the form of a blackboard or pencil and paper. Sound is better than sight for music where, although notation is crucially important, few people can look at a symphonic score and 'hear' the work. We do not know how classical Greek music sounded, because notation did not exist or failed to survive.

Until recently words rather than pictures have been the means for defining legally sanctioned and forbidden behaviour. Effective administration, legislation, and government have depended on precision in the verbal skills of parliamentary draftsmen. Now, numbers are playing a larger part, for example in environmental regulation. More recently, pictures have risen in mathematical importance, because of a revived interest in pure geometry in the form of topology and in the study of chaos and Mandelbrot fractals (Appendix B). We shall be discussing in Chapter 7 the need for picture-indexing that can enable diagrams to be used more effectively in patent law. For, despite the greater precision of numbers, words have been the means for describing intellectual property. If an invention requires a picture for its definition, the patent property tends to be weak.

The impact of the digital computer

As already mentioned, the processing and handling of words and numbers received an even greater stimulus from the introduction of computers, because these encode our standard letters and numbers into a code based just on two symbols consisting of the binary digits 1 and 0 (Fig. 2.11). This was the step that smeared out the difference between records made up of letters and those made of numerals, and encouraged us to speak instead of 'alphanumeric strings'. Some of these sequences of binary digits are the encoded data of accountancy, commerce, engineering, and science, and some the sentences of letters, books, and other texts. Vast amounts of arithmetic and algebra can be performed rapidly and great haystacks of data can be searched quickly for needles. Inventory control has suddenly become cheaper and easier, and new

---

Fig. 2.11 The binary encoding used in the digital computer.
  (a) Decimal and binary numbers.
  (b) An 8-track punched tape.
  (c) A bar code—a special form of binary coding.
(Reproduced by courtesy of the Open University.)

Computer coding

		BINARY							DECIMAL		
One-hundred-and twenty-eights	Sixty-fours	Thirty-twos	Sixteens	Eights	Fours	Twos	Units		Thousands	Hundreds	Tens	Units
2^7	2^6	2^5	2^4	2^3	2^2	2^1	2^0	10^3	10^2	10^1	10^0
							1 =				1
						1	0 =				2
						1	1 =				3
					1	0	0 =				4
					1	0	1 =				5
					1	1	0 =				6
					1	1	1 =				7
				1	0	0	0 =				8
				1	0	0	1 =				9
				1	0	1	0 =			1	0
		1	0	0	0	0	0 =			3	2
1	1	1	0	1	0	0	0 =		2	3	2

(a)

(b) TC0 (Null or Blank Tape)
TC4 (End of Tape)
FE0 (Backspace)
FE1 (Horizontal Tab)
FE2 (Newline)
FE4 (Form Feed)
ERASE
SPACE

(c) ISBN 0-00-217361-1

kinds of engineering and science possible. The only requirement is that all data must be easily reducible to a one-dimensional code, a linear sequence of standard symbols.

Initially, therefore, the arrival of the computer might have been expected to consolidate the position of records in the form of words and numbers, and the position of more sophisticated versions of the 'three Rs' in education. In fact, it has been more even-handed. The computer is so powerful a processor that it can also generate or record pictures (Fig. 2.12) by using the usual forms of scanning to convert long alphanumeric strings into pictures and shapes (Chapter 4). Computergraphics thus provides one of the ways of making and processing pictures, including three-dimensional images. In engineering, it facilitates drawing and manufacture in such a way that a draughtsman can prepare instructions for a computer-controlled machine tool, using pictures generated from standard shapes as the means for expressing his wishes and presenting his design. In illustration, it enables an unskilled person to produce good illustrative material, including perspective representations of three-dimensional objects (Chapter 5). It opens up the preparation of animated graphics, created by cartoonists like Walt Disney, to anyone with a little patience. There are special new techniques for obtaining a wide range of colour shading and effects. So just as the computer text package deskills the preparation of typing and documents, so the computer-graphics package deskills drawing. Text and pictures can be intermingled in a way that makes colour-illustrated book and newspaper production an easily learnt trade (Fig. 2.13). There remains, however, the difficulty of indexing, where words and numbers still score over images. This is the subject of Chapter 4.

The technological history of pictures

Techniques and skills for painting and sculpture have waxed and waned over the last three thousand years, but have remained in the realm of art and culture. So, despite the importance of facial expression, gesture, and movement, in identification of friends or enemies, or emotions such as pleasure or anger, or sexual desire and intention, thirty thousand years of experience of painting had not, by 1840, created a generally usable medium of communication. The good early painters were few, and, although we can identify their subjects and guess at their messages, they probably acquired little power and did not become a significant social cadre. About two hundred years ago—with the advances in science and technology—engineering drawing, and medical and other microscopic illustrations, became important adjuncts to science, medicine, and manufacture. Nevertheless, painting technology remained simple and was concerned with pens and brushes and the shade, brightness,

The technological history of pictures   31

——————————— **Computer graphics** ———————————

Fig. 2.12 Checking traffic signals throughout a town. (Reproduced by courtesy of Plessey Controls Ltd.)

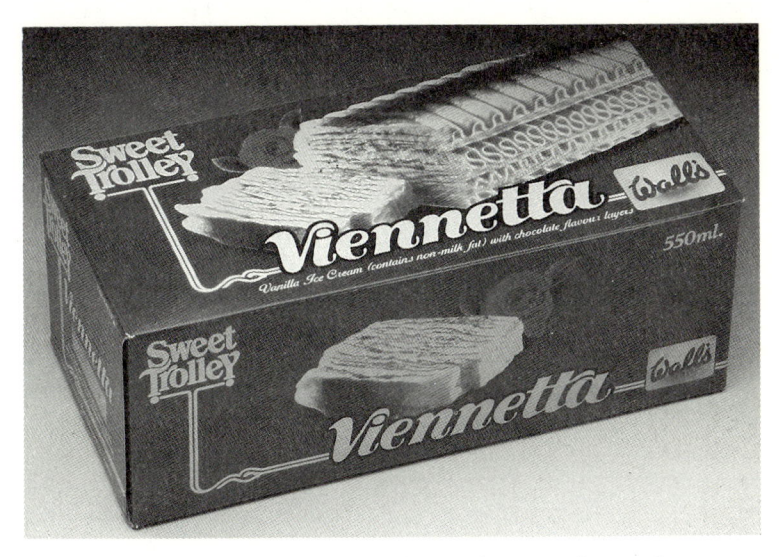

Fig. 2.13 Magazine technology applied to packaging where colour range and quality of picture are demanding. (Reproduced by courtesy of Unilever UK Central Resources Ltd.)

and durability of paints and surfaces to be painted or otherwise embellished. It also included the methods for achieving good effects with inks and paints, and the geometry of perspective.

The explosion of pictures

From about 1840 onwards, there was an explosion of pictures, as the range of methods available for picture-making began to expand. Black-and-white photography broadened the scope. The number of photographers increased as the plate or film became faster and as cameras, films, and processing became cheaper. However, photography remained the province of the professional and the specialist for most of the nineteenth century, though it helped to support the written word in the illustration of books.

Then the box Brownie made it possible for the ordinary citizen to take his own pictures. Most of these were family records of the growing-up of children. The dead could be seen by those who had never met them (as had previously been possible only for those rich enough to commission portraits). At the same time, cinematography allowed the recording of movement and events. We could see black-and-white movies of the streets of cities and of events in peace and war from 1900 onwards. Between 1919 and 1939 there was a cascade of advances in the use of pictures. In 1928, just when the silent film had become established as a new dramatic art form (open to many times the number who could go to the theatre or read books), sound-accompanied cinematography took over. Professional colour photography and cinematography became commercial activities. The television camera and receiver were invented and primitive public black-and-white television transmission was started. In 1938 the techniques of scanning in the cathode-ray-tube made possible the conversion of radar signals into pictures and maps. Scanning began to be used in the automatic recording and processing of graphical data in science and engineering.

Wartime work between 1939 and 1945 brought substantial advances in the control of colour and image persistence on cathode-ray screens. These improvements were applied in colour television after 1945, and reached general use soon after 1960. Military signal-processing and cryptography led logically to the electronic computer, the output of which was, again, displayed on the cathode-ray tube, first as numbers, then as words, and finally as numerically generated and processed pictures and diagrams. This last development became most important for engineering design and process control. Success had at long last crowned the thirty-thousand year search for technical means to use pictures in a broad range of activities. This historical development is summarized in Table 2.1.

The explosion of pictures

Meanwhile sound recording added the cheap gramophone record, the even cheaper cassette, and the compact disc. It is now possible to reduce considerably the amount of verbal writing. Curiously, however, very few people send audio-cassettes to one another. One problem is that these cannot yet be scanned quickly, or easily dipped into.

Alongside these technical developments, popular education since 1900 has provided the markets for big newspaper, book, and advertising industries, and more general affluence has financed a popular fashion trade. All of these activities called for drawing as well as photography, so commercial art arrived, developing into a world of its own creation. Much of this development was simply a popularized (or vulgarized) version of the capability that had evolved culturally; but one class of drawing—the cartoon—developed at a speed and in a way that depended on newspapers and advertising. At the same time, the spread of elementary mathematics made it increasingly possible to display quantitative information as graphs, bar charts, pie diagrams, and little pictures. In the 1930s, Walt Disney and others created the animated cartoon, which initially was too expensive for any application other than popular entertainment. Nowadays, computers make animated cartoons that are cheap enough for use in education, science teaching, and the presentation of general information. The arrival of the array processor (Chapter 4) will still further lower the cost and simplify the drawing of animated cartoons and graphics.

The amount of information that can be portrayed in colour cinematography is about a thousand times greater than that conveyed on the sound track alone. This means that the development of tape-recording, from its beginning in sound alone to the video-tape recording of sound pictures, has been a big job. The next stage, the invention of the video-disc, has required the addition of laser technology and the processing of optical signals. Now, in the 1980s, we are able so to compress data that we can assemble cheaply all kinds of pictures and words and numbers, on tape or disc. A further technical input is the liquid crystal display screen, which can replace the rather bulky cathode-ray screen with something much flatter, lighter, and more compact (though as yet considerably less versatile).

These developments have had a profound effect on telecommunications. Picture telemetry for conversation and negotiation, whereby correspondents exchange or share pictures and images across distances from a few up to hundreds of thousands of miles, is still rather dear in relation to its perceived benefits. Its use calls for education and changes in attitude. Not surprisingly, the ability to use and benefit from these new methods (such as electronic mail) has not kept pace with their almost explosive growth in the technology. More particularly, however, the fast accumulation of pictorial material requires a much more general recognition that we do not know how to index pictures so as to locate them and retrieve them systematically. This is one of the major issues addressed in this book, and is examined in Chapter 4.

Table 2.1
Historical chart to show events leading to the Picture Explosion

Historical period	Non-recorded communication	Recorded communication		
		Pictures	*Words*	*Numbers*
1 billion–30 000 years ago	Biological development of chemical communication. Sight and sound important.	?Biological development instinct?	Biological development of genetic codes (many yet unknown).	
30 000–5000 years ago	Continuing development of human gesture and voice.	Development of painting, little used in communication.	Beginning of symbolism.	
5000–2000 years ago	Feedback from symbols into richer spoken language.	Improvements in painting and sculpture.	*Early codification*	
			Formal word records. Alphabets and syntax.	Formal numbers. Start of mathematics.
0–AD 1500	Spread of Latin in European scholarly circles.	Church patronage for painting and sculpture as messages to illiterate population.	Standardization of spelling and syntax. Spread of Latin in Europe.	Arabic numerals. Spread of mensuration and mathematics.
		←——*Spread of printing*——→		
1500–AD 1840	Decline of Latin. Pride in local languages.	Pictures decline in relative importance in popular communication. Growth of pictures in biology and engineering. Telescopes and microscopes.	Spread of literacy and numeracy. Early newspapers. Beginning of modern science.	

Table 2.1—contd.

Historical period	Technology	Education
	Pictures, words, and numbers	
1840–AD 1950	Camera and its use with microscope and in science generally. Telephone, radio and television. Gramophone and records. Growth of commercial art, film and cinematography, cathode-ray-tube, scanning, radar. Electron microscope.	Near universal literacy-numeracy in European and western world. Development of newspapers, comics, and paperbacks. *Pictures lose out: words and numbers win.* Growth of science leading to new technologies and to computers.
AD 1950	General spread of television. Digital computer, C-R-T and digital computer-graphics. Scanning electron microscope. Video tapes and discs. Radio telescope. Remote sensing. Liquid crystals.	Arrival of computers. Distance learning by radio and television. Digital computers help with word and text handling. Digital computers help with development of mathematics and its applications in science and technology. Interactive video.
	←————— *An explosion of pictures* —————→	

To complicate and enrich the prospects yet further, the ability to synthesize, record, and manipulate holographic three-dimensional images is already here, and will almost certainly soon be cheap enough for general use. The meeting of ghosts instead of travel-exhausted people will be possible, but probably initially unpopular, since travel-exhaustion has become a status-symbol, and travel provides an excuse for avoiding thought and brainwork.

Communication through television, video, and cheap colour photography

So far, the common man's production of pictures has been confined to the colour stills that he can take with his automated Japanese camera, which has eliminated the need for skill in light measurement, focusing, and aperture control. Proud ancestors extensively bore their friends with colour-slides and pictures of their immortality in the form of their descendants. Progress on a job can be reported by photographs instead of by a written report. Engineers separated by large distances can collaborate by working together on an identical picture on a screen. The crew of a satellite can report to their ground station and can carry out work and resolve problems by making use of the full range of available terrestrial resources. The ideal of the instant recording of facial identity and expression has come almost overnight, so that face-to-face confrontations, with re-running for examination, or filing for posterity, are here now. Video-conferences are still expensive, but facsimile machines and simple scribble-pads can supplement audio-telephone conversations at a price more generally acceptable. Video-tapes, discs, and television are beginning to supplement books, a trend vividly expressed in a recent letter to *The Times*. The writer observed that 'the decline of the book may be more advanced than many of us feared'. His schoolboy son had stayed up late at night to watch a television presentation of Joseph Heller's *Catch-22*, which was required reading for his A-level examination. Yet 'the paperback original sits undisturbed upon the shelf'.

Obviously, however, it is the 'star' transmission of high-quality television from central points, rather than 'network' interchange, that has had the most dramatic influence on communication. Some of this is interactive, as programme 'phone-ins'. The main result has been a great increase in pictorial presentation of material to audiences of various sizes. Entertainment still leads, but the spread of colour television has made cinema-type moving pictures readily and widely available as a means for teaching, learning, and remote discussion. At present, however, the flood of pictorial capability has not led to a large volume of interactive communication by pictures.

Mainly accidentally, rather than deliberately, a very substantial proportion of the information received, understood, and remembered by children now reaches them from the television screen. 'Distance learning' mixes radio, television, books, and interactive tuition in a way that gets the best use out of all of them. Advertising by pictures is not confined to magazines, newspapers, and television. Companies with sophisticated products, such as automatic machine tools, now issue video-tapes to display their wares. One interesting area is that of machine- and car-spares identification. This is quite difficult when done with illustrated books, which are expensive to update. Car spares depots have had microfiche machines for some time, but now the presentation of descriptions and locations of parts can be done on-line to the warehouse by computer graphics. Thus an enquiry can immediately be met with a list of the places where the required spare is to be had. At the same time, the system can record its own performance and suggest improvements in the location of the stocks.

It is hard to estimate the growth in the proportion of information and skill that is acquired from television and other pictures. They are increasingly used both in teaching and research and in schools, universities, and polytechnics and industrial training schemes. Whatever the precise figure, it seems certain that the uses of pictures and words will henceforth be much more evenly balanced, and that the handicap of the picture makers and users has finally been removed.

3. Pictures, words, and numbers in childhood

Introduction

The new technological capabilities for creating and disseminating pictures call for a re-assessment of the use of pictures in relation to words and numbers. The present balance of use for any one person is a product, partly of inclination and innate skill, and partly of education and upbringing. Philosophy and methods of nurture cannot be changed quickly. This means that we need to begin to think urgently about our treatment of children in relation to their natural inclinations in the use of pictures, words, and numbers.

All of us play a part in the child's development. Commonly, the unstructured education of children by parents, other relatives, friends, and their contemporaries may be more important than the structured work done by teachers in school. It is certainly wrong to regard any rebalancing of the use of pictures as a matter for schools and teachers alone. Already the proportion of perception and learning that comes from pictures rather than words and numbers has been greatly increased by the spread of television. The precise degree of any re-adjustment will depend on the control that parents exercise over the use of the television on-off switch. Nor is it a question of 'the less television the better', or 'the more educational programmes the better'. The effect of television on the child can induce narcosis, deception, creative energy, useful learning, and much else besides. Advertising is not all a bad influence, and 'high-grade' programmes are not all good, either in their quality or their impact. Much has been written on the effects of television, some of it commercially partisan, some politically motivated, and some detached. Probably no very clear conclusions would result from a survey. Moreover, television is only one of a number of influences. Children read books with coloured illustrations and also comics. Some not only see high-quality colour photographs, but take them with cameras of a power and precision obtainable at a low cost unthinkable when their parents were young. Coloured slides are used in classroom teaching.

Colour in newspaper pictures and posters makes them better material for wall displays, in classroom or bedroom, and more significant than black-and-white in their effect on the viewer. Diagrams and animated graphics are now cheaper and more widely used. Mathematical concepts and useful arts can be taught through pictures as well as numbers: the arguments of Euclid can be alternatively encapsulated in the diagrams and word-logic of pure geometry or in the equations of algebraic geometry (Figs. 3.1, 3.2). Similar alternatives exist in the natural sciences. The correlation between crop-yield and dosage of fertilizer can be expressed as a table of digits or as a line between Cartesian co-ordinates. The new picture-technologies are enhancing the possibilities of pictorial approaches in relation to words and numbers.

We need, therefore, to begin a discussion about the development of perception and recording in the lives of children before formal education starts. We can then usefully look back at our own society fifty years ago, when words and numbers may have been at their highest point of relative influence. This experience should help in the examination of events in our own society and in our guesses about the future. In all of this, we must bear in mind the long time-lags and big inertial factors: practice may well follow a century or more behind need. So at present we are, to some extent, bringing up children on the basis of ideas and practices designed and fitted for the end of the last century, and even for times before that.

The flavour is given in an account by John Holt, in *How children learn*, of the early education of a little girl, who was allowed and encouraged to follow her inclination to draw as a central means of learning and communication, and who became highly motivated as a result. 'I wish that I could report that this tremendously productive and self-renewing process continued and grew throughout the year. It didn't. This was not because of the teacher, who was a very understanding and gentle woman, and gave these children much more time for art than would most first-grade teachers. But she was under the pressure of the curriculum, the academic lockstep, and both she and the children were under the pressure of the nervous parents, worried about their childrens' "progress"—that is, whether the Ivy League Express was running on schedule. The children began to feel, after a while, that there was no time for art, that it was not serious—and six-year-olds in school are very serious. They are also very sensitive to what adults value. They show a parent or a teacher a picture, and the adult says, in a perfunctory voice, "How nice, dear". Then they take home some idiot workbook, whose blanks have been dutifully filled in, and their parents show real joy and excitement. Soon the pictures get shoved aside by the workbooks, even though there is more learning in a good picture than in twenty workbooks. When, in later years, the children do draw pictures, they are very likely to draw them more as an escape from real life, like the war scenes of third-grade boys, or the horses that ten-year-old girls interminably draw, than as a way of getting in touch with real life.'

40   *Pictures, words, and numbers in childhood*

────────── **Mathematical concepts expressed as images** ──────────

The Chinese triangle

Pascal's Triangle

one pattern is this:

$\searrow + \swarrow$
$= \downarrow$

this works all over the triangle.

The chinese triangle is also "Pascal's triangle"

Fig. 3.1 The Chinese layout of binomial coefficients often known as Pascal's Triangle. Work of a twelve-year-old. (Reproduced by courtesy of Jane Davies.)

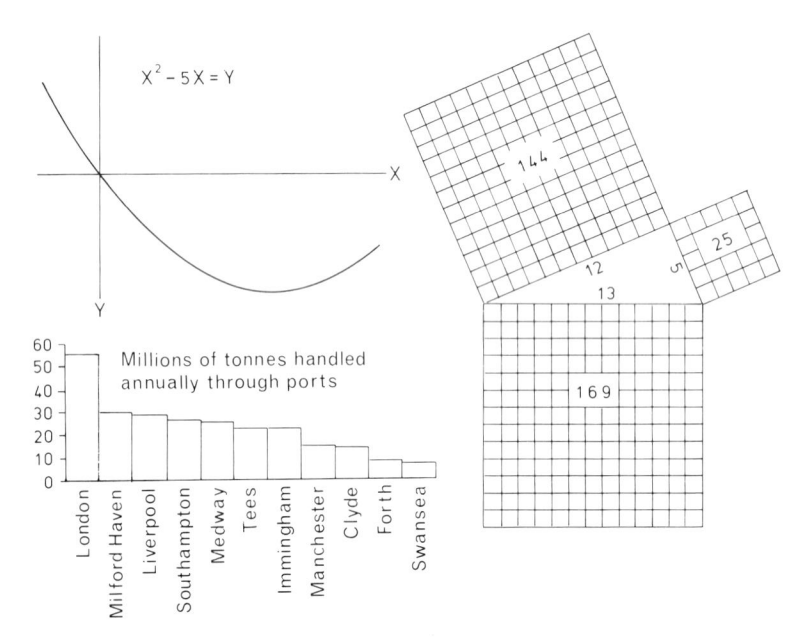

$X^2 - 5X = Y$

Fig. 3.2 Algebraic, quantitative, and geometrical concepts in diagrammatic form.

The position before formal education

Young babies rapidly change their priorities in the use of the five senses. The control systems provided by instinct are initially directed by the molecular senses of taste, smell, and touch, which guide the infants to their maternal sources of food and havens of safety. Bearing in mind the darkness and relative quietness of the womb, the rate of development in the use of sight and sound is remarkable. Photons and phonons become the carriers for the messages the child receives, and increasingly displace the earlier molecular ones. Blind children have to be taught to chew: sighted ones learn by observation of others.

Children begin to invent sounds that they like, as well as mimicking those that are found to evoke desired responses such as affection, entertainment, or the arrival of food and drink. So the early experiments in sound production initiate a development in which sound eventually becomes the dominant form of message transmission. Smiling and other facial expressions, with the support of body language, provide important back-up, but this range of transmitted visual signals is relatively limited. The story of message reception is different. During the first three years of a child's life, the range of things recognized through hearing may be wide, but recognition through vision diversifies a good deal faster. The response to things seen is to make noises which gradually become articulated into recognizable words. There is then an interesting period of two to three years during which children receive information mainly by vision, but send information by sound. These original experiments in sound are followed (when sufficient muscle control has been learnt) by experiments in drawing. At this stage children repeatedly draw what they like, or what takes their fancy, including abstract shape and colour. So begins the transmission of information by the picture.

Primitive children, not taught to read or write, sometimes develop their own semi-formalized drawings, which are used as basic pictograms. Affluent children like looking at books with pictures, and this aptitude provides a starting point for teaching and schooling. Symbols (alphabetical and numerical) attached to pictures give the names of familiar people, animals, or objects, or the number of items in the picture. Pictures are the convenient way in for words and numbers, a sort of fifth column for formalized symbols (Fig. 3.3). Inculcation has begun to bend the child's route towards the application of symbols (for it is thought to be a waste of time to leave the child to develop its own symbols of communication). Once the child can read and write, the pictures are gradually taken away, so that the balance shifts in favour of words and numbers. Certainly, in the past, many dictionaries, and most textbooks for languages, literature, and history in secondary education were almost free from pictures. It is also still true that pictures, to the

Pictures are the way in for words and numbers

Numbers

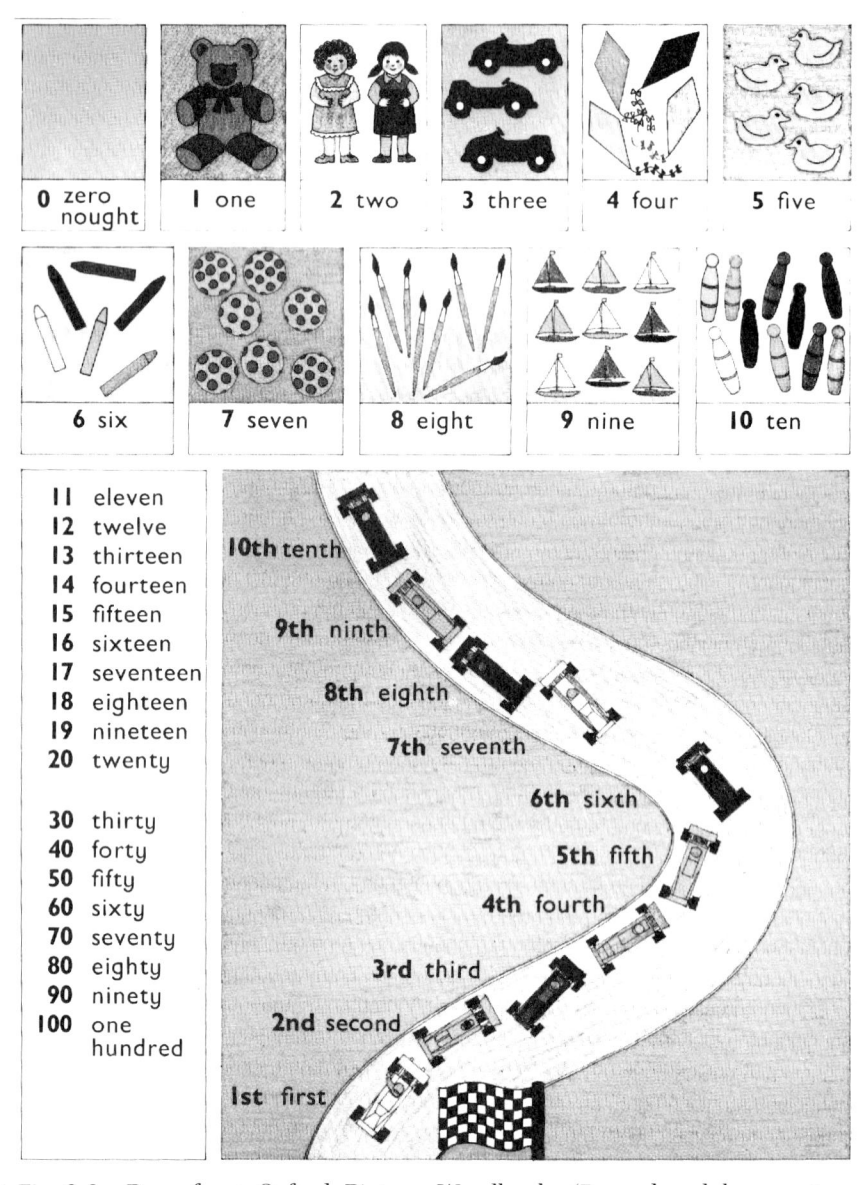

0 zero nought	1 one	2 two	3 three	4 four	5 five
6 six	7 seven	8 eight	9 nine	10 ten	

11 eleven
12 twelve
13 thirteen
14 fourteen
15 fifteen
16 sixteen
17 seventeen
18 eighteen
19 nineteen
20 twenty

30 thirty
40 forty
50 fifty
60 sixty
70 seventy
80 eighty
90 ninety
100 one hundred

10th tenth
9th ninth
8th eighth
7th seventh
6th sixth
5th fifth
4th fourth
3rd third
2nd second
1st first

Fig. 3.3 Page from Oxford Picture Wordbook. (Reproduced by courtesy Oxford University Press.)

upwardly-mobile parent anxious to see his own attitudes mirrored in his children, are amongst the things he believes that St Paul set aside when he came to noble adulthood. Illustrated comics are tolerated; but proud parents are pleased when children are seen reading the sparsely-illustrated quality newspapers. Only the dullards are expected to require the pictorial stimulation of the tabloids, where the images are those of sex and violence.

The position in education up to fifty years ago

There are good pragmatic reasons why educators and parents have arrived at the present preference for words and numbers among the skills that able children should acquire. Returning to the historical trends described in the last chapter, literacy and numeracy in the Middle Ages were skills needed for the priesthood and useful in trade. The feudal aristocracy required different skills and attainments, many connected with military equitation. From Tudor times onward, feudalism decayed and printing became progressively more important. Learning and letters became less dependent on clerical use and patronage, and more closely connected with the growing mercantile classes, banking, and the processes of government. Consequently, although the ambitious could still prosper through the military prowess that built and maintained empires, more and more rose to eminence and authority through the use of words and numbers. The new professions of civil and mechanical engineering began with rule-of-thumb, but quickly became dependent on calculation. With increasing numbers of people collaborating, success also depended on the writing and talking necessary for good administration and management. Engineering drawing and design were also needed, but in Britain those so skilled tended to be on tap but not on top. Engineers were more highly regarded elsewhere (especially in Germany, France, the USA, and Japan), and it may prove that this factor will be a valuable advantage in the growth of pictorial skills in those countries. In a word, the literate and numerate did well; science and teaching became important as high-class infrastructure, which further encouraged the rise of the clerks and the mathematicians.

The teachers, called upon to demonstrate their own prowess and utility, found that the easiest way to do so was by the production of unchallengeable score-sheets for accurate spelling and arithmetic (Figs. 3.4, 3.5). Literary skills were a little more difficult to mark unchallengeably, and artistic abilities were the most difficult of all. Consequently, the imperatives of social and commercial purpose on the one hand, and demagogic authority on the other, were both met by the growth of that proportion of the population able to spell, read, and count. The 'Ivy League Express' referred to by John Holt was the vehicle of

44 *Pictures, words, and numbers in childhood*

———————— **Unchallengeable marking: computation** ————————

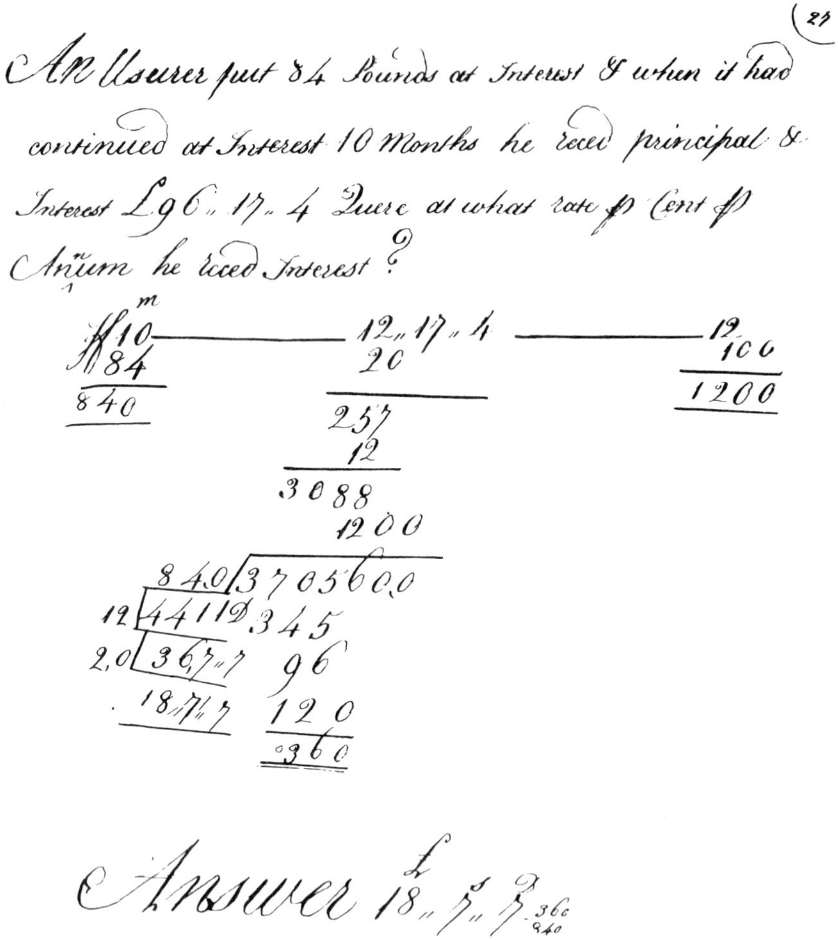

An Usurer put 84 Pounds at Interest & when it had continued at Interest 10 Months he rec'd principal & Interest £96..17..4 Quere at what rate p (ent p) Anum he rec'd Interest?

Answer 18..5..5

Fig. 3.4 Mathematical exercise from John Spurell's school book, dated Dec. 8 1746, Hindolveston, Norfolk. It has a familiar look for pupils of two hundred years later. (Reproduced by courtesy of the Education Library, University of Liverpool.)

Pictorial expression in the primary school

Plate 2 A primary school classroom 1987. Pictorial work being produced by children and covering the walls. (Reproduced by courtesy of Delyth Edwards. Borthyn School, Ruthin, North Wales.) See p. 46.

Pictorial exhibition in the secondary school

Plate 3 Wall display in a biology laboratory (contrast with Plate 2). (Reproduced by courtesy of Shorefields Community Comprehensive School, Dingle, Toxteth, Liverpool and University of Liverpool.) See p. 47.

Unchallengeable marking: spelling

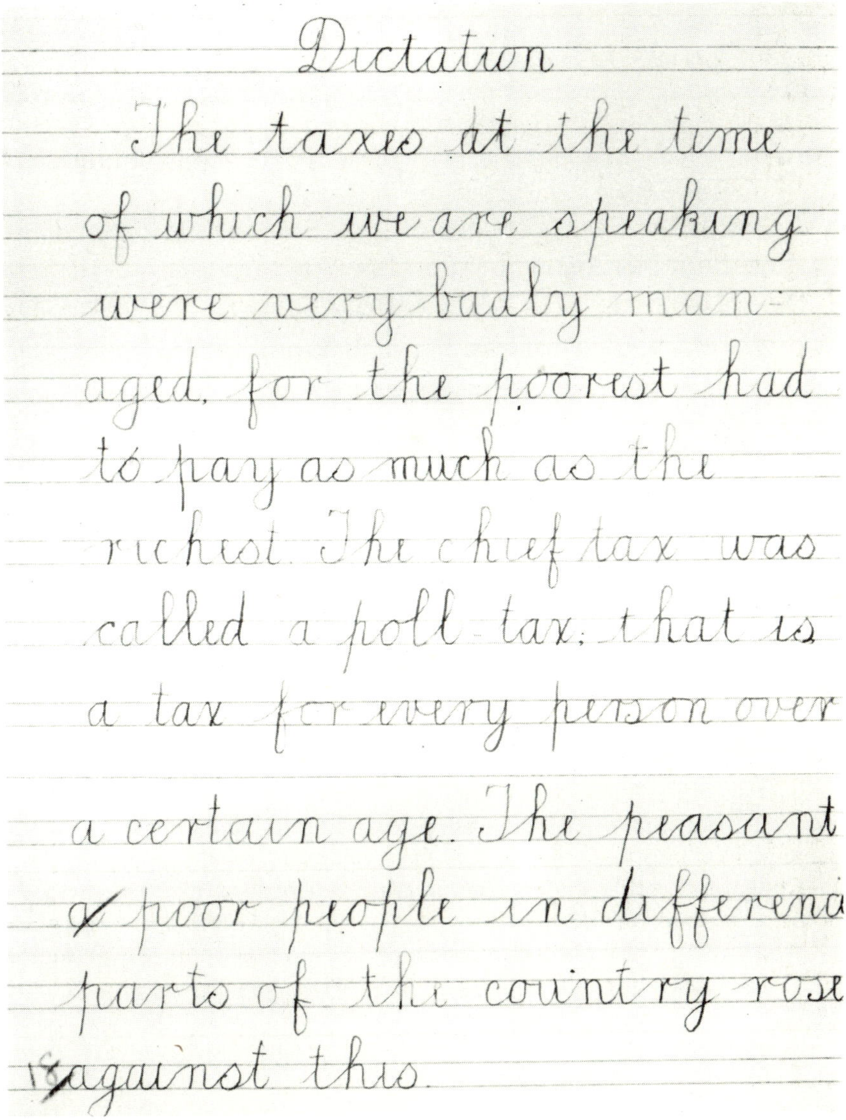

> *Dictation*
>
> The taxes at the time of which we are speaking were very badly managed, for the poorest had to pay as much as the richest. The chief tax was called a poll-tax; that is a tax for every person over a certain age. The peasant & poor people in different parts of the country rose against this.

Fig. 3.5 Dictation exercise from Ethel Harrison's school book, dated April 11 1905, when she was ten years old. (Reproduced by courtesy of A. Carter and the Education Library, University of Liverpool.)

economic growth, and not just a holiday train. Reading and counting became known as 'hard' subjects, and pictorial skills as 'soft'. These concepts became institutionalized because they were descriptors of beguiling and convenient simplicity. They were also difficult to displace by the more tedious processes of logical argument. The new situations and the new needs of science, trade, engineering, and government, required pictures as well as words for their description and discussion; but 'soft' pictures could not gain equality with 'hard' alphanumeric strings. The proper balance was difficult to establish.

Pictures and words in education now

In the last chapter, attention was drawn to the extraordinary speed of the arrival, during the last three decades, of half a dozen capabilities that make the production of pictures much easier and picture-based communication a current reality. We all know that children watch a lot of television, and that they see video-displays of computer games. Those who have access to Acorn and Spectrum computers at home can play such games themselves. Others play computer games in amusement arcades. We are aware of the use of television by the Open University and of the existence in schools of television sets and computers that have pictorial programmes. (These subjects return in Chapter 7.) How do the children themselves assess pictures, words, and numbers, and how are their priorities formed at school?

First and foremost, there is one group of teachers in all countries who cannot escape association with the child's transition to a life increasingly dominated by written words and numbers. These teachers work in primary schools. In affluent countries they receive television-aware five-year-olds who want to draw pictures as well as watch television. Yet parents continue to call for progress in words and numbers, and the teachers have a strong sense of their own importance and vocation as the purveyors of the three Rs. Nevertheless, it is in the primary schools that the flame of the picture has been kept alive. Project-work yields wall displays which provide an important way of uniting the disparate abilities of their pupils. Although pictures as 'art' are difficult to mark, as illustrations of a project about the Medieval Village, or the Big Cats, or Crossing the Ocean, they are an essential accompaniment to the words (Plate 2). Everyone cuts them out or draws new ones, and all are pleased when their offerings are included in the class display. Parents see them on Open Night, and accept them as valuable aids to the writing and spelling to which they are still committed. Moreover, in spite of the hypnosis and narcosis of too much television, children do see many programmes from which they take in much of their database as pictures. That which enters the child's mind as a picture can motivate and assist the playback of the information by the child, most naturally as a drawing. Slides, computer graphics, and video-tapes are now much used in well-equipped classrooms (of which we need more). Educational suppliers, as well as the broadcasting and commercial video companies and agencies, sell teaching filmstrips, slides, and videos as well as books. This spectacular increase in the child's pictorial intake is followed only slowly by a growth in his pictorial output.

Further evidence of the power and importance of the child's pictorial perceptions and playback is furnished by the psychologists, psychiatrists, and child-guidance workers. Many of these ask children to draw pictures as an essential part of their diagnostic procedures. These pictures can provide a channel to the child's worries and concerns that may not have been articulated at the conscious level. Literature is full of stories of images that transcended the available words and sentences, and dreams are recalled in pictorial form. Joseph's dream of the seven lean and seven fat kine is a good example.

Psychologists who study memory are also clear that pictorial recall has special importance to many people. Some remember complex routes through cities by recalling the buildings on the corners at which they must turn. Others remember the position on the page where a particular item of information first came to their notice. Ways of remembering can be improved by learning, and it is said that pictorial association was more extensively used in a deliberate manner when, centuries ago, few people could read or write. Features in a familiar room or church were remembered, and the items to be recalled were associated with the features of the familiar environment. The story-tellers of ancient Greece and Rome are believed to have used this method for remembering long narratives. The technique is sometimes called 'Pelmanism'. It is alleged that an ingenious undergraduate at Cambridge once codified all of his knowledge by the superimposition on standard diagrams of an elaborate stained glass window in the examination room, and did well as a result. He certainly deserved to, but why do the rest of us make so little use of this kind of pictorialization?

At the secondary and tertiary stages of education, we are still, to some extent, stuck in our 1900-based ways. The examination system calls for only a small amount of picture, important as it is in practical work. In history, language, and literature there is little scope for pictorial expression. Geographers, on the other hand, have always used pictures such as maps or sections, and, more recently, diagrams and graphs. In secondary schools pictures tend to be provided for the children as sources of information, and are used far less as means of expression (Plate 3). One has only to watch the Christmas Lectures televised from the Royal Institution to realize the importance of the use of pictures, an in-house tradition that goes back many decades. There are important ideograms in chemistry and physics, as we shall see in Chapter 5, and naturalists, biologists, geologists, and doctors have always been keen users of good illustration. It may even be that the convenience of photography and video is doing some harm by eliminating the critical faculties of the scientific illustrator: to draw a thing is to get to know it intimately. The importance of pictures and diagrams in some fields may be one reason why, in certain countries such as Britain, the best brains have been deflected from picture-linked engineering and even from science—an attitude as expensive as it is absurd.

Few who are upwardly mobile take examinations in art, although some who are already affluent feel able to do so, and to study art at the more advanced and expensive schools, alongside music and drama. Those doing so are probably well-assured of the connections needed for a good job. University lecturers use more slide material than before, and photography is an integral part of much scientific research and teaching. But are the university visual-aids departments beseiged by eager customers? For some of the highly educated, pictures are either an art-form to be looked at in a scholarly fashion at exhibitions, or television entertainment about which they sometimes express a measure of apology or shame. 'I had nothing much to do last night, and felt pretty tired, so I watched the box for a few minutes.' There then follows some intellectual excuse, explaining that such box-watching is aimed at the scholarly study of one's fellow-men, or the inadequacy of politicians, or the use of language.

Yet the educational potential of television is immense (Fig. 3.6). As is usual, innovation is concentrated where people are tackling new educational tasks and opportunities. One such area is that of 'distance learning'. This takes several forms. At one extreme it is the means whereby educational resources can be provided in remote communities, such as Indian villages or small isolated settlements in Australia. Teachers are still crucial, but radio and satellite television can supply material that is better presented than it would be in books that are anyway too expensive. The transistor radio is accepted in poor and primitive communities as legitimate and useful entertainment, so that it is an aid rather than an obstacle to educational, vocational, or cultural objectives. At the other extreme, agencies such as the British Open University (one of our most important 'world firsts'), are playing a major part in 'continuing education'—the attack on the obsolescence of the skills and ideas of older people—and in part-time higher education. It is the means whereby those who have missed out or dropped out can join the educational stream. The Open University is, for practical reasons, doing the educational research and development that the full-time, traditional system finds it difficult to regard as important. In particular, the Open Unversity is having to identify which parts of higher education are suitable for radio and television, how the scarce resources of interactive teaching (summer schools, telephone tutors, and correspondence) can best be deployed, and what mix can best be fitted in with the fatigue and demands of job and family. Then there is the new Open College,

New aids to learning

Fig. 3.6 A new kind of reference material, the Doomsday Video, as part of the school library. This is an interactive system whereby the student can select from a very large database, and choose her own route through the information. (Reproduced by courtesy of Acorn Computers Ltd.)

Fig. 3.7 Interactive video instruction in office skills in a high-street learning centre. The student responds to the choices by touching the screen. (Reproduced by courtesy of Alfred Marks Bureau Ltd.)

which aims to provide vocational training at a sub-degree level through distance learning guided through television and video (Fig. 8.2, p. 135). Other important frontier educational areas are training on the job, and specialist instruction in such matters as safety and quality control and the new high-street interactive-video learning-centres for office skills (Fig. 3.7). Interactive video is discussed in Chapter 8.

What is emerging from present practice and research is a first perception of the range of good mixes of pictures, words, and numbers. Rarely has any one of the three no part to play, but the richness of the options is bewildering: we have stumbled into an Aladdin's cave. Not only is the mix different for different subjects, but it varies for different teachers, communicators, pupils, and receivers. Some can manage numbers better than others. Sometimes numbers, especially in economics and social affairs, have such large and unquantifiable margins of error that their very use can create a spurious and dangerous impression of precision. Monthly figures for the nation's balance of payments can lead to mistaken changes of policy. Sometimes words can precisely capture and epitomize a whole continent of analysis and thought. 'Matter and energy are interconvertible, at a fixed exchange rate' means more to an accountant than does Einstein's relation between energy and mass expressed as '$E = mc^2$' (where c is the velocity of light). Sometimes one needs the picture, a caption, and a remark in a balloon (as in a cartoon). Sometimes the fourth dimension—movement—shows up important information that a flat or stationary image fails to convey. It is the management of this mix that is the central subject of this book. The inadequacy of our present information is surprising, but not as surprising as the lack of general concern about the whole business.

Pictures and words in future education

Many questions face us. Are pen and paper obsolete? Are the new recording methods going to reveal a need to elaborate, enrich, rationalize, or otherwise develop spoken language? Will pictures make body language more important, and may that create a new interest in dance, drama, and mime? Will educators need to teach the practice, detection, and understanding of deception and distortion (as well as honest communication) through the new powerful media such as television? Most particularly—and this is the subject of Chapter 4—how serious is our inability to index and recall pictures with the same facility to which we are accustomed when using words in dictionaries and encyclopedias?

Speculation is important as a basis for the discussion on which we can base policy. There will, however, be a simultaneous need for discipline (to make the discussion manageable) and freedom (to bring in all the important topics, some of which we cannot yet perceive). At this stage, it seems most useful to enunciate some simple, and perhaps obvious, suggestions:

1. Educational change will be slow and patchy. The mainstream will take a century or more to change, even if the rate is greater than that in the recent past (which was itself greater than ever before). In some places, as already indicated, there will be opportunities to move faster; those who do this will be the object of jealousy, envy, and resentment from the practitioners of orthodoxy. This will be especially true if the new ways enjoy economic success; then there will be a danger of backward-looking dogma and the persecution of heretics. Moreover, traditionalists will sometimes be right, and reformers wrong. Expect trouble and violence.

2. Present methods and equipment will continue to be useful. The videotape or disc will not replace the book, which is convenient, compact, and good for browsing. Moreover the very limitations of a book make it a creator of economical and critical thought. Books are the poetry of the future. Pens, pencils, and paper, and sculpture in marble will remain for similar reasons. (Sir Roy Strong suggested that we may be becoming two nations, one whose reading is restricted to the tabloid press and the telephone directory (and which is reliant on television for news, information, and entertainment), the other for which the literate dailies and books are a source of information and delight. 'As illiteracy proliferates' he wrote in *The Independent*, 'those who value literacy become increasingly aware that the ability to read is what sets them apart'.). One important matter is autonomy: one can do algebra or geometry, or write poems, by scratching with a nail on the wall of a prison cell. The new modern methods are dependent on the supply of batteries or of mains electricity, and on availability of spares and people who can do repairs. Even now, electric power-points are not available everywhere, and natural catastrophes or industrial strife can lead to an interruption of the power supply. Stores of batteries cannot be taken for granted, even with the assistance of solar power. Whether there will be more or less reliability remains to be seen. Perhaps we should encourage the pursuit of autonomous and extra-reliable equipment. Someone needs to invent the pedal-operated computer/radio/cassette-player; perhaps this will be built in a school as an educational aid.

3. Present affluence may hinder educational reform as much as it helps. Good cooking tends to result from moderate poverty and the need for ingenuity. It may be that the need to overcome social obstacles to the use of the new capabilities will stimulate the successful search for solutions. The man-machine interface is still likely to be a central problem, which may be made more difficult by the latest equipment. Black-and-white photography is sometimes preferred when the emphasis is to be on composition, and simple

pictures are sometimes best in displaying clear messages about people and their habits. Elaborate equipment can become an end in itself, and produce results that are technically superb but lacking in content.

4. The young may be the teachers as well as the pupils. It used to be true that children often used the telephone with more ease and effectiveness than their more knowledgeable and experienced seniors, and the same seems to be happening now with computers. Teachers attest to the speed with which children acquire the necessary skills in school.

5. Disability may become a more or a less serious problem. On the one hand, a dyslexic may find new ways of communicating, and may be better at these than his peers. Already, with the help of computers and videos, some physically helpless paraplegics have produced literature of great value. Periods of change lay emphasis on the flexible use of the brain, and those with less mental flexibility may be more socially disabled than heretofore.

Nearly all the other issues to which we now turn carry educational consequences. We shall, therefore, return to the question of future education in our last chapter.

4. Pictures, words, and numbers: sorting and indexing

Background

We must now face an unsolved problem of fundamental importance and great difficulty: the sorting and indexing of pictures. To understand the situation we must spend a little time examining how lexicographers, librarians, and scholars sort words, how accountants, scientists, and engineers sort numbers, and how computers sort either or both. These tasks we can perform with precision, because we understand all of the processes, having designed them as part of our alphabetic and alphanumeric techniques. We must next observe that the mammalian brain in general, and the human brain in particular, can recognize and sort pictures at a level of sophistication that is quite unattainable by computer-based methods. We can then note what it is that the computer can do in this field, and guess what it is that the brain does that we would like the computer to emulate. Unfortunately, the advances that have been made in the understanding of brain anatomy and physiology, although great, are not yet sufficient for our present needs.

First, however, it is necessary to indicate why this sorting and indexing is so important. Action based on experience and knowledge requires that we assemble large numbers of different observations in a search for regularities. These regularities can then be used to formulate theories (or models), and thus to make predictions that test them. The basis of these predictions has changed.

At one time, our experience was assembled somewhat irrationally, in the form of statements about Gods and their likes and dislikes. Certain days, it was believed, were unfavourable for the undertaking of journeys or business, because the Gods were angry with those who broke the (incomprehensible) rules. Conclusions could be reached by examining the entrails of sacrificed animals. Even if experience seemed to contradict the rules, it was often inadvisable to set up a challenge, because the rules as they stood were the stock-in-trade and power-base of a strong priesthood which had the authority to punish heresy.

In the last three centuries, however, we have had great success in collecting related observations systematically without inhibition or prejudgement. These accumulated observations have then been searched for empirical correlations to use as a basis for models or theories that could, in turn, be employed predictively. These models, with their observational bases, and the tests of their forecasts, have collectively been called 'science'. In our examination of the natural world, this approach has been more plausible, comprehensible, and successful than the earlier use of divine models. Of Newton's Laws, Medawar has said that 'it is no longer necessary to record the fall of every apple'. Given knowledge of the height of the branch and the mass of the planet on which the tree is growing (that is, the value for gravity), then the time of fall (in the absence of an atmosphere) can be exactly calculated. In the presence of an atmosphere, such as ours, we also need the models of the aerodynamicists, or an approximation such as Stoke's Law, to correct the calculation for air resistance (the viscosity of the fluid). Thus equipped, and perhaps with attention to Einstein's relativistic modifications of the model, we can calculate the results of free fall, and guide men to the moon, or a probe to Uranus or to Halley's Comet. On a quite different scale in the laboratory, we can, on the same basis calculate accurately the physical dimensions of fine particles in terms of their rates of fall through a fluid of known viscosity.

The assembly of information for formulating new models (or for designing the equipment to use them) requires the ability to sort out the observations that are relevant from the mass of data that is not. With alphabetic language we can do this, for a search under several keywords will rapidly separate wheat from chaff. Before the computer, the search was manual and very laborious. Nevertheless, the scientific and engineering literature was indexed alphabetically, so that conditions for the search did in fact exist. With the computer, the search can be rapid and unerring, provided the initial indexing has been accurate enough.

Apart from scientists and engineers, many others wanted to search reliably and quickly. An unfamiliar word is encountered, perhaps in a foreign language: look it up in the dictionary. Information is wanted for an essay: look the subject up in the encyclopedia. An important case, to which we shall return, is that of a patent search. Scrutiny of the invention in relation to the 'prior art' requires diligent and effective searching of the published literature. Another enormous area of searching and sorting embraces telephone switching and connection. The caller may be connected to the recipient by a single line. Alternatively, and more economically, a single line may be used simultaneously by a number of callers to an equivalent number of recipients. In this case, messages between each pair are separated by being carried on different frequencies or by occupying specified, regular, time slots (Fig. 4.1). In either case, the destination of a call is identified by a rationally chosen number properly catalogued and related to information about the recipient, as in a

Sorting information

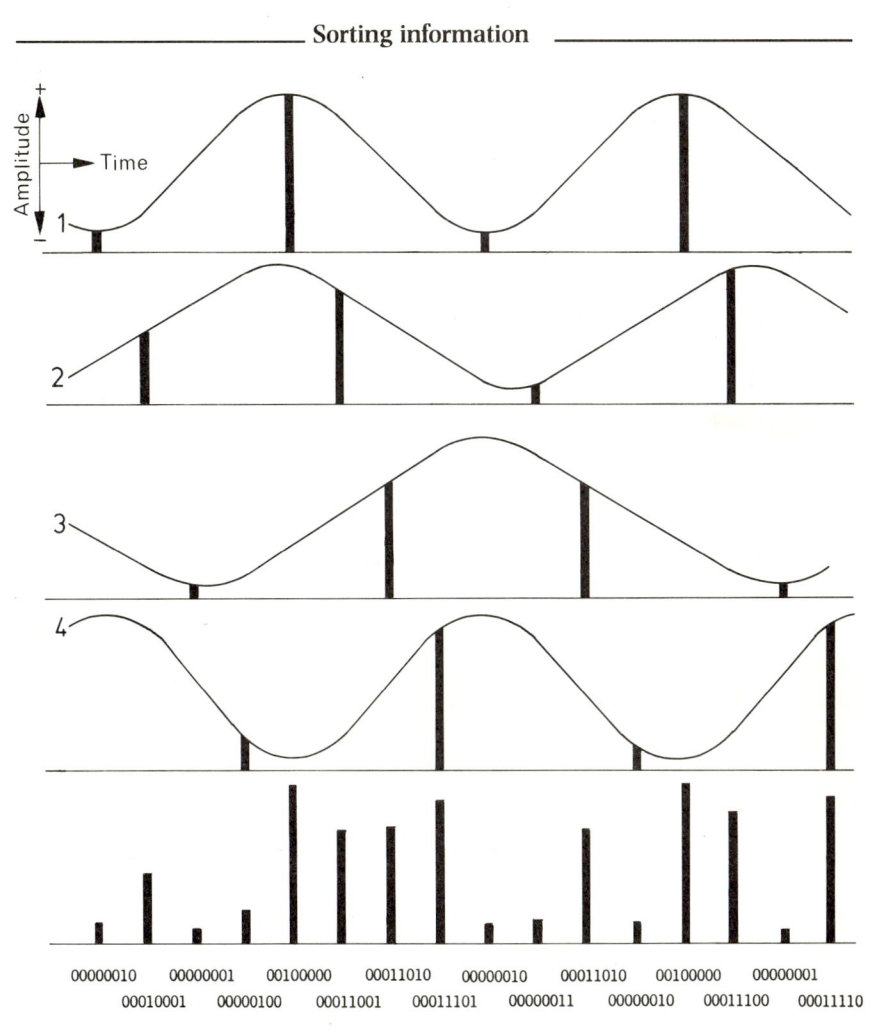

Fig. 4.1 The simultaneous use by four callers of one telephone line. The sound waves in the four channels are transformed into analogous current oscillations, which are sampled at regular intervals (the vertical lines). Information about amplitude and frequency at each sample is encoded as a binary number (digitized), and then transmitted as a pulse within its own time slot (along bottom line), so separate conversations are carried by each fourth pulse. (After *Reliable digital communication*, Hobsons Science Support Series.)

telephone directory. In quite another way, at a personal level, our own eyes and ears are continually sorting the things in which we are interested from those in which we are not.

Sorting words: further manipulation of the alphabet

The lexicographer or his customer sorts words by visually picking out the letters, deciding where the word will be placed in an alphabetical hierarchy, and then inserting it or finding it. Blind people have to use braille or sound as the basis. The computer is blind and deaf, but creates, and responds to, pulsed sequences of electric current. These pulses are the result of on/off switching according to a code based on numbers to the power of two, that is, the binary code (Chapter 2 and Fig. 2.11, p. 29). This code uses only two symbols, 0 and 1, and these can be directly reflected as the two switch positions. This simple system enables us to encode the ten decimal arabic numerals 0–9 as the computer alphanumeric strings of the binary code 000, 001, 010, 011, 100, 101, 110, 111, 1000, and 1001. The machine can then read arabic numerals, translate them into its own binary alphabet, do binary arithmetic of any desired complexity very quickly, translate back into arabic numerals, and print out the result. Similarly, other binary alphanumeric strings are used to encode the phonetic alphabet, after which the computer can search for and sort words just as it can search for and sort numbers.

Circuitry can be devised to search for, and assemble, all words beginning with the letter 'c', or all entries containing the word 'guitar'. Sequences of dates can be arranged in chronological order, or sequences of employees in order of salary. The computer can easily identify all lawyers who have brown hair and ride a bicycle, in case such a lawyer is needed. A special form of binary coding—bar-coding—is used for the books in a lending library, and assists searching and disciplined loan. The same bar-coding simplifies payment at supermarket checkouts, and simultaneously feeds back data for stock control (Fig. 2.11, p. 29).

The computer can also be given the rules for writing verse, and allowed to use a built-in dictionary to select words so as to arrange them in a poem that scans and rhymes. Accountancy numbers can be sorted out and added up in categories for budgeting and checking purposes. Nevertheless, the computer cannot do word arithmetic as effectively as it does number arithmetic, because we have not devised word-syntax in as disciplined a form as our number-syntax. But in a matter of seconds it can look at every word in a ten-page text, search the built-in computer dictionary to see if the word is there, and print out the text and positions of all rogue words. Some of these will be typing errors needing correction, and others will be proper names or technical terms that don't happen to be in the dictionary. Any of these can be added, at will, to the (now personalized) dictionary, so that they will not be thrown up next time. By putting an 'x', or other convenient symbol, before any word that should appear in the index, the computer can be made to generate the index. Computer syntax can then be employed to arrange the index—or any list—in alphabetical order.

Sorting pictures: the absence of picture alphabets and syntax

Before we look at the reasons why the sorting and indexing of pictures have assumed such importance, it is necessary to think a little more closely about the processes of scanning and synthesis in television cameras and digital computers. We shall then be able to see what is needed if computers are to be able to search and sort pictures or images as effectively as they can sort words and numbers. This is the problem of pattern-recognition referred to in Chapter 1.

Image transformation involves the conversion of light images into electric signals that can be used by a computer. A television camera (on Earth or in a satellite) records pictures through a process of scanning. It breaks down the moving scene into a sequence of stills, each of which it reviews, not as a whole, but by analysing it as a mosaic of lines of spots. The camera sweeps or scans the entire mosaic of spots, line by line, about fifty times a second, transforming the light energy of each spot into an electric charge. These charges can then be used to generate radio signals, the intensities of which vary directly with the intensities of light received. Computers can be employed to digitize this analogue information in a manner akin to the system of sampling shown in Fig. 4.1. The computer divides the rectangular picture into small regions called picture elements or pixels (Fig. 4.2). Circuitry synchronizes the pixels with the picture scanning. At each pixel the signal is converted to a digital value which lies in a range of numbers corelated with a scale of greyness from white to black (for example, arabic numbers 0 for white and 255 for black as in Figs. 4.2, 4.3). This coded information, called raw data, is transformed into radio waves when transmitted.

At the receiver, the reverse processes take place, and the picture is reconstructed (or synthesized) on the screen, spot by spot, line by line. The digital computer can, therefore, by using binary code, help in the transmission of pictures, or, since it has a memory, store them, or, synthesize new ones—a kind of 'painting by numbers'.

Wherever the pictorial information moves, in whatever form, it is transported via a carrier. Just as pressure-pulses in the air form the carrier for the sound of our voices or music, so light waves are the carrier for visual

58 *Pictures, words, and numbers: sorting and indexing*

――――――――――― Scanning and digitizing a picture ―――――――――――

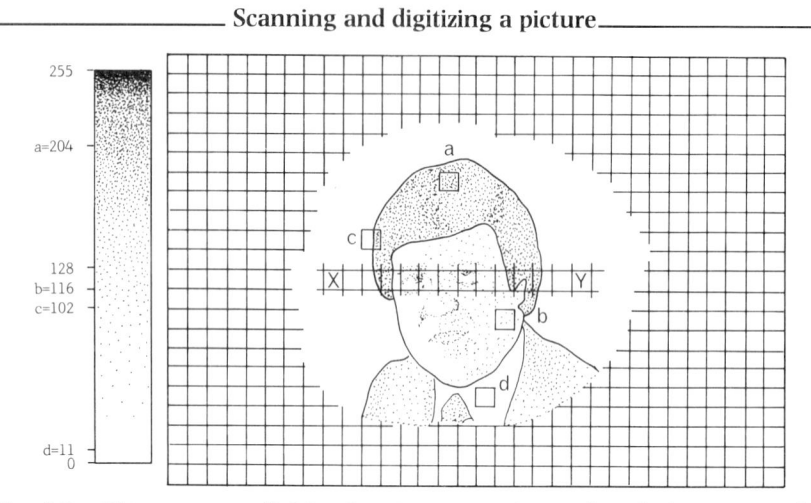

Fig. 4.2 The computer divides the picture into lines of pixels (squares in this diagram). Each pixel is expressed as a number according to its position on a numerical scale of greyness. Note that pixel c, which includes some light background and some dark hair, has an averaged value which is similar to that of pixel b on the face.

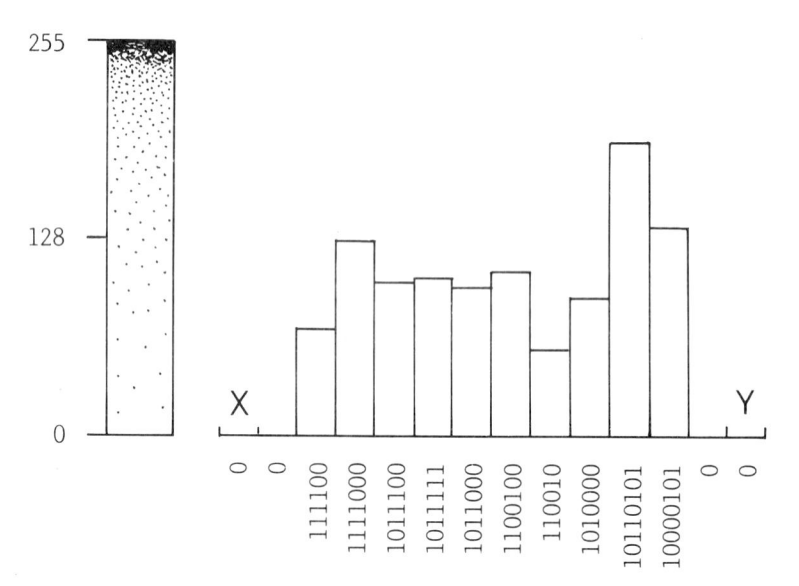

Fig. 4.3 The pixels in line X–Y in Fig. 4.2 are scanned for their grey-scale values, which are then encoded into binary numbers. The final number will include the pixel address (row/column).

Sorting pictures: the absence of picture alphabets and syntax 59

images. The sound pulses, or waves, may be treated, alternatively, as streams of particles called phonons, the light waves as streams of photons, and the oscillating electric currents (also with wave-like form) as streams of electrons (Table 4.1).

Table 4.1
Carriers and their merits

Carrier	Sorting	Storage	Transmitting	Coding
Phonons (sound)	—	—	—	★
Electrons (electricity)	★★★★	★★	★★	★
Photons (light)	★	★★★★	★★★★	★★
Molecules (biological)	?	★	★	★★★★★

Superior	=	★★★★★	Fair	=	★
Excellent	=	★★★★	Poor	=	—
Not bad	=	★★			

In speed of transmission, photons and electrons handsomely beat anything that nerve or intracellular processes can manage. They win also in density of particle transmission. In computers they approach the standard of the brain for storage. For coding biological molecules sweep ahead.

We now return to the problem of the sorting and indexing of pictures. These processes involve searching, and the need for this is not far to seek. Remote sensor and camera data from satellites or aeroplanes are continually being telemetered or brought back to base in very large quantities (Fig. 4.4). These raw data commonly arrive already translated into digital form. On arrival this information is stored mainly on magnetic tape (which can deteriorate in a few years). Contained somewhere in this digitized material are pictures of deposits of minerals, oil, and waste; traffic jams; disasters that have not yet been notified; crimes about to be committed; missile silos; and so forth (Plate 4, Fig. 4.5). How can these items be identified and retrieved so that appropriate action may be taken? Our problem is that the elements of pictures, such as individual ships or aircraft, or buildings of a particular type, or clouds of a particular description, are not described by the picture equivalent of the uniquely defined word or sentence. They are built up by numbers. There is an absence of standardization. The same house gives rise to quite different

60   *Pictures, words, and numbers: sorting and indexing*

————————— Satellites provide the familiar weather map —————————

Fig. 4.4 Remote sensor image (infra-red) of the cloud formation in a depression on 31.1.1988. East Anglia can be seen in the clearing to the right of centre, and the Bay of Genoa towards the bottom right. Such pictures are built up, line by line, on the earth station's receiving screens. Satellite NOAA-9. (Reproduced by courtesy of the University of Dundee.)

sequences of binary numbers according to the state of the weather, the angle from which it is viewed, the time of day, its distance from the sensor or camera, and various other factors. Similarly, the same digital numbers may describe quite different objects (Fig. 4.6). Somehow or other, the human brain is able to span all these complications, and to say of a series of photographs, 'These are pictures of Keats's house in Hampstead, and those are not' (or 'that is cloud and that is sun'). At present the abstracting, and analysis, of information from the computer images, which are built from raw data, is done by the human brain. This is exemplified by the technician who wrote 'I spend my time analysing the 35 000 photographs crews have brought back from 24 missions'. Such abstracting would be greatly assisted if computers could recognize shapes (or patterns) with the efficiency that they can recognize letters or numbers. NASA has enormous programmes on such 'image recognition', but has not got very far.

It is now appropriate to look at what the brain can do, and to consider to what extent the computer might be adapted to perform similar tasks.

The brain versus the computer: contrasting capabilities

Both the brain and the computer have truly remarkable capabilities, but they excel in different tasks. The computer can perform calculations, or store and retrieve precisely defined data, at speeds far in excess of those attained in the brain (Table 4.1). On the other hand, the brain of a child of less than two can distinguish its mother from its aunt, or indeed from a complex background of chairs and cupboards and a hat and coat hanging from a hook. These tasks are beyond the computer's competence. Nor can a computer identify the whole of an object from the sight of a part of it; yet a child easily recognizes father by his beard alone. The computer is no good at spotting associations between seemingly unrelated pieces of information and deriving generalizations, for example, the notion that pictures of a house, a train, and an aeroplane all have in common the possession of windows. A not dissimilar problem arises in attempts to design computers to translate from one language into another. 'Manslaughter' might be equated with 'funny human' simply because, again, the computer has not appreciated the significance of context.

The brain scans and files information away differently from the computer. The image of a particular acquaintance can be stored in the brain and retrieved ten years later. The retrieved image can be compared with the current image of someone seen in the street, with hair of a different colour, clothes of a new fashion, and signs of ten years' wear and tear. It is then somehow or other possible to decide positively that the stored and current images are, or are not, of the same person. What is more, the same scan (completed within a second or

62   *Pictures, words, and numbers: sorting and indexing*

_____ The computer cannot recognize features of pictures _____

Fig. 4.5

Fig. 4.6

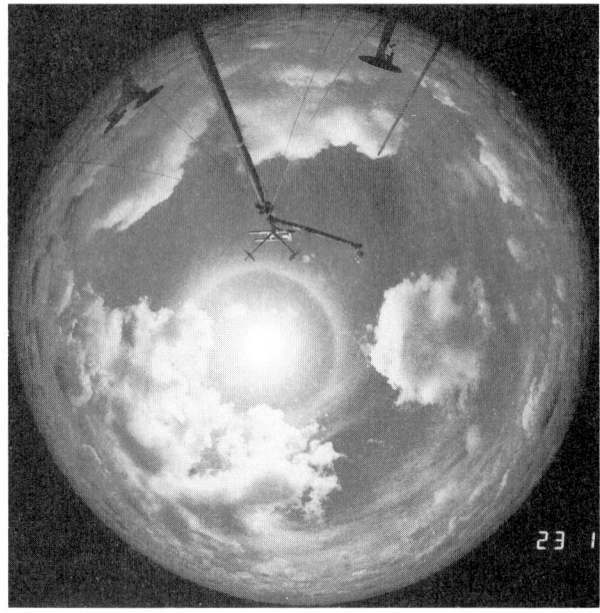

two) can cover the question of the person's happiness or misery, the characteristics of the walk or cant of the head, the consideration of whether the degree of ageing is more or less than would be expected, whether the subject has gone down in the world or risen to greater prosperity, and much about his or her relationship with companions. 'Gracious, that must be Beryl! How she has aged!'

By these standards, the methodology and capability of the analogue or digital television camera and the digital computer in pattern-recognition are poor indeed. The computer's laborious process of dissecting the picture or frame into readings of colour and density of the hundred thousand or so pixels, which are the basis of the record, is slow and uninformative. The manner of the recording fails to distinguish the component sub-images (trees, flowers, people) that our eyes and brains can recognize. Thus the computer cannot at all easily say, 'this picture contains a boat, a lady with a parasol, an island, and a swan', let alone give any details about any of these items.

The computer's incapacity relative to the brain does not result from any failure in its speed of sorting, processing, or transmitting of binary numbers, or in the storage capacity of its memory. It was noted earlier that the carriers could be electrons or photons. As we shall see later, in Chapter 6, nerve messages, which are carried via chemical reactions, move much more slowly than the electrical or light signals of the computer and its sensory equipment. The storage capacities of computers are rapidly surpassing what appears to be the brain's storage, good though that is (Table 4.1, p. 59). It is the brain's

---

Fig. 4.5 An image of the Thames Barrier, with the Victoria and Albert Docks to the north. The image, built of pixels each representing an area of 7 m × 7 m, shows the silt in the river. The docks have clear, dark water, the river has sediment throughout, and the turbulence behind the piers has caused greater amounts of sediment to be picked up from the river bed. These are features which the human brain can detect. The computer can build and display this picture from the digits which represent the light intensity at each pixel, but it cannot recognize the differing amounts of sediment in the river, nor make sense of the pattern of silt movement. Neither can it recognize the railway nor a road. (Reproduced by courtesy of Hunting Technical Services Ltd.)

Fig. 4.6 A photograph taken with a fish-eye lens on board ship. This picture could be digitized, stored, and displayed by a computer. The digits would represent the light intensity at each pixel, and the same number would represent those parts of the sun and those parts of the clouds which reflect the same light intensities. So the computer could not differentiate between clouds and sun, nor distinguish the haze which is apparent to the human eye. (Reproduced by courtesy of Ann Henderson-Sellers.)

cryptology that is so remarkable. Between the 'letters' into which it subdivides pictures and the totality of the information that the person is receiving, there must be several levels of rationally related subinformation, corresponding to words, sentences, and syntactical connections. It is tempting to think that the brain files away much bigger assemblies of information than the computer in the form of coded shapes of biological molecules (a matter considered in Chapters 5 and 6). Chemistry may make it possible for the brain to encode external pictures by the use of its own internal 'pictures', made up of three-dimensional molecular orientations and positions. While the computer acts upon received information serially, carrying out sequences of operations successively, the brain seems to carry out several operations at the same time along parallel pathways. In so doing it probably makes use of the virtually infinite number of different possible assemblages of interconnected neurons (Fig. 6.20, p. 108). On this basis it can learn, and make use of the learnt information to improve the efficiency of its operations. Attempts have been made to develop computers that carry out many operations simultaneously (in parallel rather than in series), as in the array processor; but these have yet to achieve the efficiency of the brain where the different operations relate and react to one another while they are being carried out (Appendix A).

If this view of the brain's operations is correct, then mechanized and automated sorting and searching of pictorial information at the level of competence of the brain must await scientific progress of a fundamental kind. What do we do meanwhile? Is it best to work away at systems that involve the use of the brain as part of the sorting process, rather as we use the brain as part of the semi-automated factory? Should we try to develop 'fuzzy' computer-sorting that will begin to make useful comparisons of similar but not identical images on the basis of new protocols? (A comparison of doorways, bridges, and chairs should reveal that a feature they have in common is the arch.) Should we work harder on the alphanumeric labelling and keywording of pictures, and employ large numbers of people to watch television programmes and video-tapes, to abstract them verbally, and then to process the abstracts by the usual methods? How much work should we do on the understanding of the brain? Will this benefit from intensive effort, or do such endeavours correspond to the attempt to produce a baby in one month instead of nine by making nine women pregnant simultaneously?

Lines of attack on the sorting and indexing of pictures

There are two kinds of approach to the intellectually very difficult problem of mimicking (or paralleling or out-performing) the brain's ability to handle visual images. One is the incremental method; do the simpler things that we can, with our rapidly improving digital computers, perhaps aided by re-born analogue machines. The other is to conduct physiological, psychological, and mathematical research on the brain itself, from time to time checking our ideas by making synthetic machines. It is probably best to begin with the incremental and severely pragmatic approach. In future, we may be able to perform quite new tasks by the sophisticated automatic processing of picture data. The analytical video-computer may produce even more new possibilities than has the word-computer. For the present, it will do no harm to give thought to the tasks that we know can be done, either with pictures or words, and to design the more immediate of our research programmes round these perceptions. This approach will help us to think about the new balance between pictures and words, and where the educational priorities may lie.

In the first place, we know which instructional tasks are particularly suitable for each of the two media. The Open University, making use of both media, shows no sign of walking away from words into television. Its courses rely on books of printed words, embellished with illustrative pictures and diagrams, all of which are verbally indexed. Most of its interactive teaching relies on correspondence and telephone conversations—again in words. However, for two weeks in the year there is a summer school when the interaction is face-to-face, and, throughout the year, there is a sequence of television programmes. What is happening is that the television slots, which initially were largely devoted to lectures (perhaps thought to be more compelling than books as a way of presenting words) are increasingly used for 'outside broadcasts' of material which is largely non-verbal (although, of course, there is a commentary). Geology, engineering, biology, plays, dance, and music are a few of the subjects especially suitable. For the understanding of a machine, its working, repair, and maintenance, a few pictures are worth many thousands of words. Study of the winning of materials and energy from the earth's crust is greatly helped by video-tapes of mines, oil rigs, windmills, wave-power machines, and animated graphics to show what goes on. Most of these items are too remote or expensive to visit, and the dimension of motion makes the tapes more useful than the book illustration. Sometimes the addition of the third dimension is particularly useful—as in the inspection of the strucures of enzymes to discover where their action is centred (Chapter 6) or of the three-dimensional pattern in which crystals of quartz, felspar, and mica are arranged in a granite so as to determine its crystallization history (Fig. 8.9, p. 147).

The preparation of such teaching material is expensive if it all has to be filmed on site. By now there are in existence large collections of films and tapes, which, if edited and catalogued (using the brain plus the computer), could provide most if not all of the instructional material. But where is it all? Who owns it? How greedy are they for royalties? How good is it and how well will it copy? Television producers face the same problems. Where is there a sunset over a stormy sea? Who has pictures of flying fish or condors? At present, it is the memories of the old hands that provide most of the answers. The provision of comfortable places for coffee breaks, where the young can consult the old, might be the best way of spending money. Already some major collections of pictures have had to be thrown out because there were no funds to pay for cataloguing.

Accordingly, there is a strong case for the expenditure of human effort in alphanumeric cataloguing. Such work, if well led, is likely to permit comparisons of different routines for scrutiny and description, which, in due course, may help in the design both of new computer methods and of the missing picture-syntax. The BBC Film and Video Library has started just such a project on transmitted programmes (but not on the material from which they have been culled).

Another necessary activity is the archiving of engineering drawings, designs, and design-based routines for numerically-controlled machine tools. Here, the syntactical Tower of Babel is there for all to see and curse. If engineers in different companies or design offices are seeking to collaborate in a common design exercise, and want to do so on screen, iteratively, they will probably find that they cannot. Such collaboration is needed, for example, between the sub-contractors for a particular component and those who are going to assemble it into a bigger system (for example, the car component-makers and car-assemblers), These people are probably using different computer-aided design packages, or using the same package according to different routines. Working together requires either translation, or the mutual learning of a graphic-exchange system. Such systems, with quite expensive translator routines, exist and are being developed. How are we to index the design of wheels so that we do not have to redesign a wheel that has already been invented? Within the big and sophisticated companies, a start has been made on such problems, and it will be important to watch this work so as to make use of any ideas that have broad applicability.

A smaller range of indexing, but with pictures that have not been computer-generated, is being worked out by art historians. There is a system called ICONCLASS, in which the index elements are those that would occur naturally to art lovers. The first series of categories begins with 'God' (where there is an awkward problem of the hierarchy of different Gods), and then works down through 'angels', 'saints', 'princes of the Church' (which church?), and so on. At a more mundane level, university libraries are beginning to acquire video-tapes, and to wonder how to arrange them on the shelves, and how to make up for the complete absence of indexes—pictorial or alphanumeric—within the tapes. The British Library is one of the agencies with the difficult task of indexing satellite-based remote-sensing data, of which there is a large and growing un-indexed collection.

In Chapter 5 we shall write of one community, the chemists, who can describe simple molecular structures either as two-dimensional ideograms or as linear, word-like strings, known as the Wiswesser Line Notation or WLN. There are computer packages for translating between the two systems (but there is the usual dificulty about machine and software specificity). If the library in your own laboratory has wanted badly enough to do the translation and has bought the right hardware and packages, you will be able to search for two-dimensional structures without learning WLN. If not, you may not. The important point is that chemistry contains Rosetta Stones linking ideograms with alphanumeric strings, the study of which may inspire others.

At the other end of the investigation—that of fundamental research—workers in artificial intelligence are trying to get to grips with the kind of subject that we are now addressing. Donald Michie and Rory Johnston's book, *The creative computer* (1984) presents a very clear and optimistic view of the present state of thought. Others are trying to build 'molecular computers', using the uniquely precise three-dimensional structures of molecules instead of electrons as information-carriers, thus mimicking the brain in a primitive way.

The strengths and weaknesses of picture and word methods

We must sum up at this stage by making a preliminary judgement on the basis of what we can already do, and what we think may be possible within a decade or so. Further ahead, there lies the problem set by Aldous Huxley in his wonderfully perceptive book of the early 1930s *Brave new world*. We have now replaced the simpler use of human muscle—for digging, pile-driving, drilling and so on—by machines. We have also replaced the use of the human brain for controlling drills, lathes, and other equipment, by programmed electric control-circuitry. We have begun to add sensors and self-optimizing feedback to the automated control systems, thus taking over the more tedious use of human sight, hearing, touch, and smell. The computer can now beat nearly everyone at chess, by an approach based on very dedicated caution and stupidity. (This is a nasty shock for us, because we thought that chess was not a game for the cautious and stupid.) If we can replace all the combinations of brain and muscle—such as the co-ordination of hand and eye in table tennis, and the making of inventions that the human race finds it wants—what role is left for man? Since, as social animals, we are not very successful, will it be time for us to bow out and hand over our technology to the socially more sophisticated insects, perhaps hoping selfishly that it will eliminate them too?

Bearing such possibilities in mind, back to the balancing of pictures and words. Words are best when we have a good model, like Newton's Laws or Maxwell's equations. When conjured with by Shakespeare or Dante, they trigger the brain very delectably and economically, and poetry is the *multum in parvo* of intellectual abstraction. So we use words when we have arrived, and we use numbers on many kinds of journey.

Pictures alone (including cinema), or in combination with words as 'talkies', plays, and opera, represent another kind of arrival. We suggest that pictures, or picture-word combinations, are better than words alone for most of the process of travelling hopefully. This is part—but not the whole—of the reason why most people are more afraid of blindness than deafness. So much of our manoeuvring—for whatever purpose—depends on the relative position and appearance of things, that the picture in these circumstances really is worth a thousand words. In the next two chapters we shall see many examples of this idea.

5. Pictures and words in engineering, physics, and chemistry

Background

Scientists and engineers have always been enthusiastic and capable users of sketches, drawings, and coloured pictures over a wide range of sophistication. Their media have varied from the backs of old envelopes, drawing-office paper, and the cathode-ray screen, to liquid crystal displays, the finest-grained photographic film, and holography good enough to substitute for reality in public exhibitions of works of art such as sculpture. Their publications, teaching, and discussions almost always employ diagrams and photographs (Figs. 5.1, 5.2, 5.3, 5.4). Accordingly, it is helpful to ask how they have reacted to the new, enhanced, pictorial capabilities. The answer is 'patchily'. Some have welcomed greater skill with open arms. Others have testily enquired what was so wonderful about something that they have always done. One point of note is that engineering drawing, like typing, has been de-skilled by the arrival of the computer. The engineering draughtsman previously had a long apprenticeship. It now takes only a month or so for him to learn how to use computer-aided design to create the instructions for making a complex item involving the cutting and grinding of metal. To be sure, this is only a part of the skill that a draughtsman has to learn, but the traditional requirement of 'ability to use the drawing board', as a key criterion of the good engineer, has been eroded (Plates 5, 7, 8). Coincidentally, the other requirement, a 'skill in cutting metal', is also obsolescent, now that other materials are just as important. Moulded shapes can be produced quickly and economically, so that cutting them out is, by comparison, wasteful and expensive.

One of the key features of the relationship between pictures and words—the immediacy of the impact of the picture—was illustrated by the tragic events of the launch disaster of the Challenger shuttle on January 28, 1986. The NASA commentator, watching the precise outputs of key numbers and words on the screen, sounded puzzled, saying, 'We have lost the downlink: we have a major malfunction', while the millions of the watching public, on site or in front of

Types of drawings used by scientists and engineers

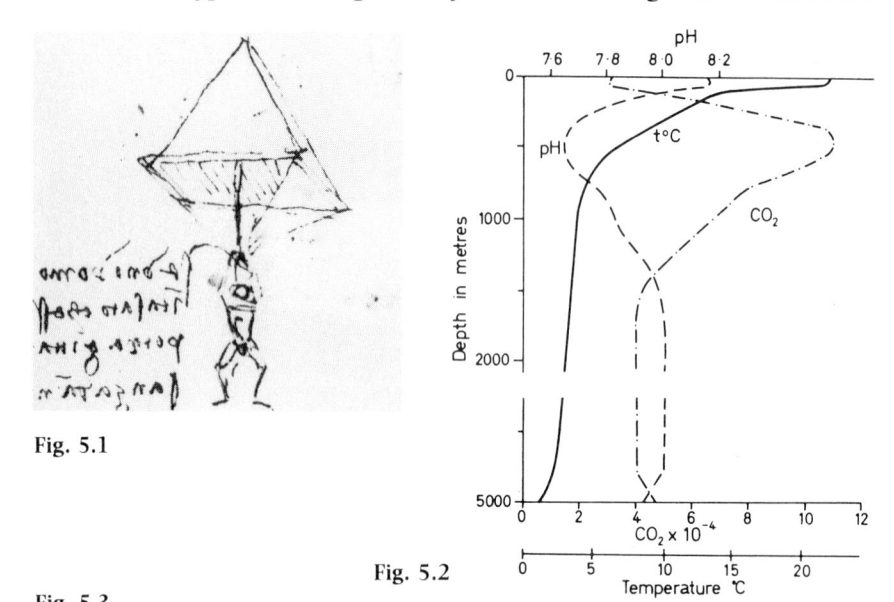

Fig. 5.1

Fig. 5.2

Fig. 5.3

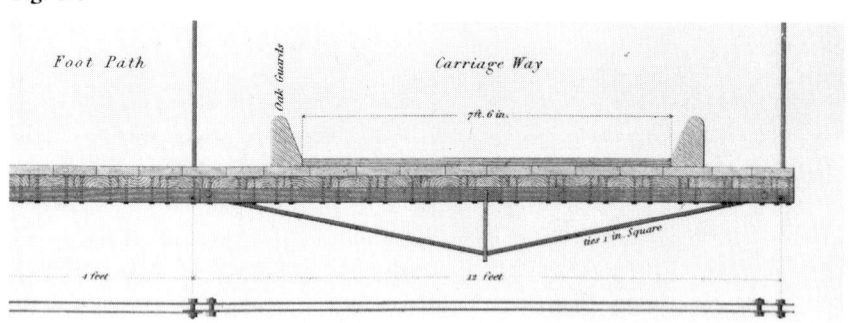

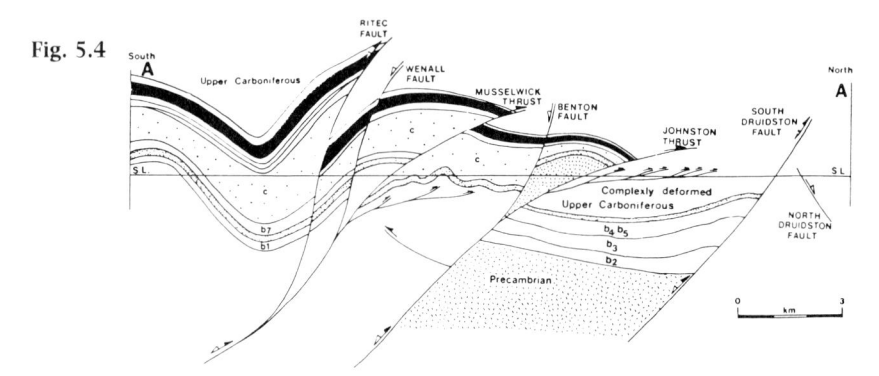

Fig. 5.4

television sets, had seen that the shuttle was already totally destroyed by fire and explosion. This case, where 'a picture was worth a thousand words', should be contrasted with the handling of the data from the downlink of the crippled Apollo XIII in the early 1970s. For nursing the malfunctioning capsule back to an improvised earth re-entry, pictures were of far less importance than numbers. The data from the vehicle had to be processed furiously over hours and days by the combined power of the ground computers, so as to work out and decide between the rescue options.

The scientist or engineer is concerned with two issues in communication involving pictures and words. The first of these is to do with the conduct of his job and his communications with his direct or indirect professional interlocutors. These may talk different professional languages, as well as different national languages. Each discipline has its own jargon. While the organic chemist would have no difficulty in understanding that 'the two carboxyl groups on the thalic acid can chelate divalent zinc', he might well be stumped by the comment of the petroleum geologist that 'stratigraphic traps are related to fault blocks and halokinetic movements'—and vice versa. The second concern is with the presentation of work to outsiders who know little or nothing about the professional method or detail, but need to know in order to allocate funds or to make use of or sell the resulting benefits (Fig. 5.5).

Pictures in computer-aided engineering

Engineers are familiar with two-dimensional drawings and have the mental ability to visualize solid objects from plans and elevations and vice versa. Alongside these abilities lies a strong desire to 'see the item as it is' from any

---

Fig. 5.1 Sketch for a parachute by Leonardo da Vinci, 1485. (Reproduced by courtesy of the Trustees of the Science Museum [London].)

Fig. 5.2 An abstraction derived from a large number of measurements. A graph showing the vertical distribution of pH, temperature and Pco_2 in waters of the Atlantic. (Reproduced from *Carbonate sediments and their diagenesis* by courtesy of Elsevier Publishing Co. ©.)

Fig. 5.3 Skilled engineering drawing for Menai Bridge by Thomas Telford. Bridge opened 1826. (Reproduced by courtesy of the Trustees of the Science Museum [London].)

Fig. 5.4 A reconstruction. A geological vertical section interpreted from surface observations in Pembrokeshire. (Reproduced from 'Inversion tectonics in S.W. Dyfed' in *Proceedings of Geologists' Association*, Vol. 98.)

72   *Pictures and words in engineering, physics, and chemistry*

───────────────── **Pictures help to find paymasters** ─────────────────

Fig. 5.5 A 100 mph collision between a diesel train and a nuclear flask was organized to demonstrate confidence in the strength of the nuclear container. The diesel was destroyed; the nuclear flask unscathed. (Reproduced by courtesy of the Central Electricity Generating Board.)

─────────── **Vulnerable to an uncalculated combination of events** ───────────

Fig. 5.6 The Tacoma Narrows Bridge. High winds caused the bridge to sway, undulate, and then collapse in 1940. (Reproduced by courtesy of Associated Press.)

Pictures in computer-aided engineering

angle, and to perceive modifications easily. They also want to discern how designed components will fit together, before or while looking at equations and graphs. Very often such visualization can expose a gap or basic error in the design. Computers allow all of this to be done quickly. A component or a system that has been calculated to be satisfactory for the standard duties of resistance to compression, tension, and twist, may look vulnerable to some combination of the three that has not been worked out. Two examples may be mentioned—both alas, in the field of the explanation of disaster after the event.

First, the suspension bridge over the Tacoma Narrows in eastern Canada failed spectacularly because of the perturbation of wind patterns by objects upwind. It so happened that someone filmed the incident, so that generations of engineers have seen this film, with the immediacy of the picture, as a cautionary tale (Fig. 5.6). Secondly, and more recently, several members of the same class of very large freight-carrying ship have mysteriously failed just forward of the bridge structure. In such cases, courts of law are more likely to be convinced by moving pictures of models than by calculations. There are many examples of disasters that have been averted by watchful stressmen who have an eye for the strengths and weaknesses of objects of many kinds. Unfortunately, the accident that didn't happen impresses nobody: the one that did happen impresses everybody.

Chemists and physicists, who are making fluids such as petrochemicals or unshaped solids such as raw plastic chips ('stuff' rather than 'things'), are not so devoted to photographs, movies, and drawings. They can usually convince their peers, and sometimes their customers, by using words and numbers, with graphs and bar-charts. It is as well for them to recognize that they need to use realistic pictures, if possible coloured and moving, when they are presenting their ideas and proposals to engineers. The latter feel, with good cause, that the chemists and physicists are talking pidgin-language if pictures are missing. Those who make 'stuff' need to talk cordially and clearly with those who convert the 'stuff' to 'things', and should study the language problems that arise in this dialogue. Pictures can help to overcome these difficulties of communication.

The computer can be used to speed up the production of engineering drawings (Plate 5) and to *begin* the process of their indexing. A firm or a drawing office might decide to index all clocks, or dial displays, or welded joints of particular dimensions or pattern. The computer can be made to do this automatically. It can take the labour out of labelling and facilitate the rearrangement of drawings in the interest of clarity. The work can be stored on disc or film, making it more accessible both locally and remotely. Using telecommunications, several draughtsmen in different continents can work together on the same picture. Probably the most advanced area of two-dimensional (or 'two-and-a-half dimensional') drawing is that concerned with microelectronics, in the design of microcircuits containing up to a million or more elements (Plate 6). Even when multilayer chips become general, the Cartesian regularity of circuits must keep drawings to the fore.

The major contribution of the computer to engineering, however, lies in its ability to move automatically between two and three dimensions, and to change flat projections into solid pictures that can then be turned, twisted, stretched, compressed, struck, chipped, and otherwise maltreated, in the manner of real life (Plates 5, 7, 8). The engineer with a really good system can look at designs and ideas so as to use his amplified experience of the Law of Inanimate Malice, much of which is unquantified, and which is enshrined by aviators in the phrase 'There's always bloody something'. A further boon is the ability to make immediate animated pictures, so that the observer can watch the design put through a sequence of styles of motion or distortion. This capability is assisted by the so-called 'array processors', which are computers consisting of many little and cheap computers in a regular pattern. These can all be fed with the data for the picture of an object. Each can then move it on a little from the position occupied by its neighbour, so that at a stroke, a still can be converted into the frame of a movie, which can equally quickly be assembled and shown. (It is clearly ridiculous to be describing this process in words: this book should really be a video-tape.) Altogether then, it is not surprising that engineers are often people of few words, or that people of few words (for example, Scotsmen) become engineers.

Most engineers are in the business of taking general principles and models, put together by others, and using them for practical jobs in specific instances. They move from the simple and the understood to the complex and only partially understood. Moreover, the need for communication varies. Although the solutions of the particular problems encountered in building a particular aeroplane may be applicable elsewhere, often they are not. The language of engineers can be expected to move further and further into the realm of pictures as a means of dealing with the particular. Alongside the doctors and biologists, they may turn out to be the most handicapped by the inability of the computer to match the brain in the indexing and sorting of pictures. Indeed, restricted as we are here by the constraints of a book (with only words, numbers, and still pictures), we may expect trouble in our attempts to describe the present state of the art of image-processing as it is performed by the digital computer.

Finally, it is worthy of note that movies, in three either real or simulated dimensions, are turning out to be of great value also to chemists and physicists, so that traditional linguistic barriers between disciplines are tending to disappear from natural causes.

Pictures in the physical sciences: chemistry

There is a view, formulated by the educational psychologist Liam Hudson, that chemists are particularly well-balanced in the use of two languages—those of numeracy and literacy. They use numbers where the sums are manageable (as in spectroscopy or thermodynamics), and concepts where the sums are more trouble than they are worth (as in the fussier applications of wave mechanics to the description of the geometry of molecules). There is a similar balance between the chemists' use of words and of pictures, and this is a situation in which chemists—by taking a very broad view of their skills—can help with the pioneering. For chemists are among the most sophisticated users of pictograms and ideograms. These range from the pictograms of the computer-generated protein configuration, which show the actual three-dimension arrangement of atoms in a molecule (increasingly used in biological chemistry for the study of enzyme action and other phenomena), to the ideograms of the two-dimensional chemical structural formulae.

These pictures are related, not only to the words that describe the chemists' models and theories, but also to systems of computer words (more precisely alphanumeric strings) that have been invented to assist in searching for and sorting chemical information. The main system of this kind is that devised by Wiswesser in the 1960s. It transforms the chemists' pictorial structures (which are often difficult to sort by computer) into linear alphanumeric strings that can easily be computer-sorted. Some examples of Wismesser and conventional two-dimensional descriptors are shown in Table 5.1.

Few chemists use, or even know about, Wiswesser notation. (They reasonably devote far more of their time and interest to the synthesis, study, and measurement of the properties of molecules than to searching the literature). With existing methods and disciplines, the requirements are two-fold. On the one hand, there is a need for computer retrieval by linearizing the description of molecules unambiguously. On the other hand, chemists require pictorial representation that summarizes chemical properties. The Wiswesser formulae say nothing immediate about properties, and must be translated into two-dimensions (or three) before such messages emerge. This accounts for some of the lack of interest in Wiswesser. Table 5.1 is, to most chemists, like a Rosetta Stone that links the use of a familiar language with that of a strange one. (In this case it is the language of the ideogram that is familiar and the alphanumeric one that is unfamiliar.) The use of Wiswesser notation for searching and sorting information means that, for most chemists, computer retrieval requires special software and hardware for translating the descriptive structures of the ideogram into Wiswesser. For this reason, searching often requires the intervention of library professionals. Indeed, the chemical library

Table 5.1
Chemists' languages and ideograms

Name	Two-dimensional ideogram	Wiswesser Notation
Phenol		QR
Pyridine		T6NJ
Tetrahydrofuran		T5OTJ
Quinazoline		T66 BN DNJ
2-cyclopentyl-1-[(3-hydroxy cyclohexyl)thio]-1-[methoxy methyl) methylsulphonio] propane		L6TJASYS1&1O1 &Y1&-AL5TJ&Q

is an awkward mixture of the computer-indexed book library and the corner-shop video-library.

The human memory has no such need for different languages for description, sorting, and searching. As we have mentioned, it is possible to enter a crowded hall and *simultaneously* to recognize an individual last seen ten or more years previously *and* to identify many characteristics all in a few seconds. In other words, the tags for retrieval and the tags for description can be recognized at once by the same memory device. We have also suggested in Chapter 4 that the mammalian memory must be chemically based, the information entering and being accessible all in one slab, for example, as a protein molecule configuration. This, if true, must encourage us to attempt the direct retrieval of descriptive structures, so as to machine-mimic the versatility of the human brain. Studies on this subject have been pursued using creatures with simpler memories such as snails and molluscs. The disorder resulting

Remote sensing images

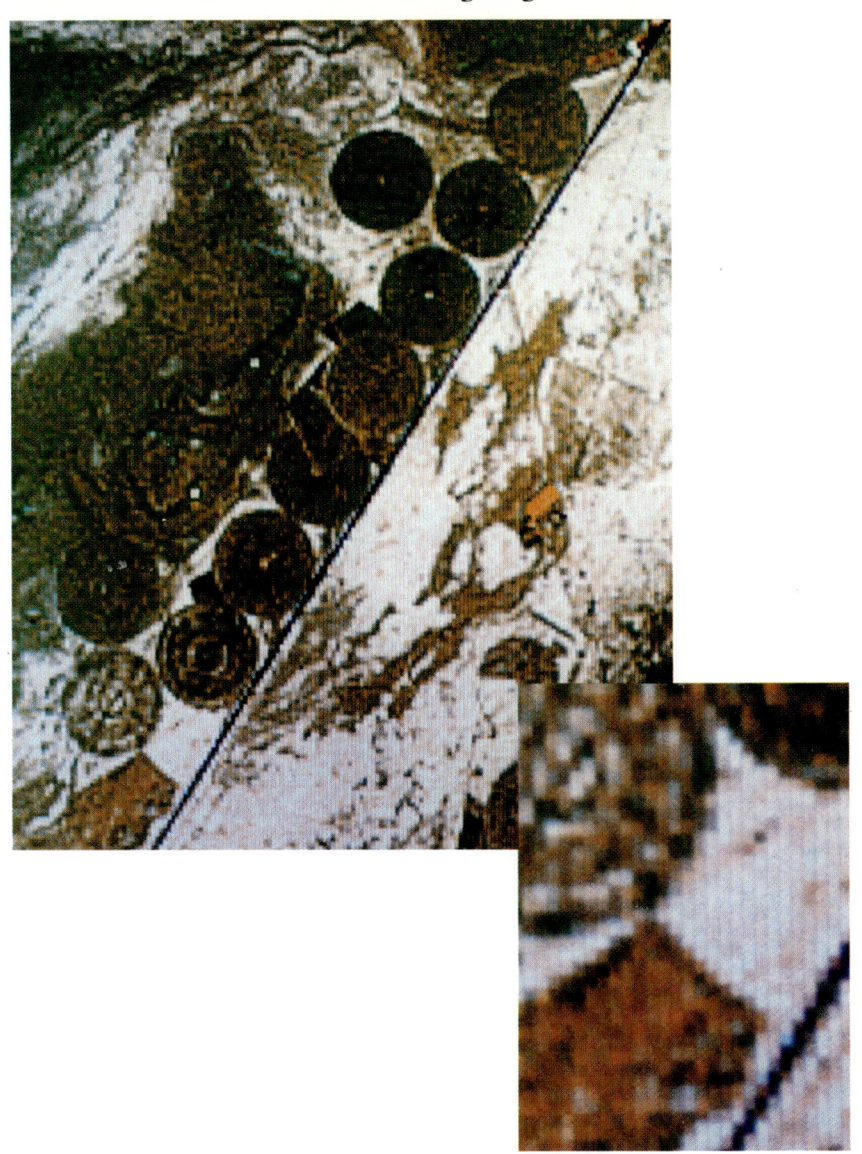

Plate 4 Pivot farms in Saudi Arabia; circular irrigation units about 1 km in diameter. An image derived from a satellite sensor at a height of 705 km. The inset shows the mosaic of pixels from which the picture is built. The picture first appears on the screen in black and white as in Fig. 4.3. A computer is programmed to colour the image—different shades of grey being given a colour code. These colours are false colours, and can be varied by the computer operator to enhance the features to be studied. In this picture green represents vegetation and white a sand surface. (Reproduced by kind permission of Clyde Surveys Ltd, Maidenhead, SL6 8BU, England.) See p. 59.

Computer-aided design in engineering

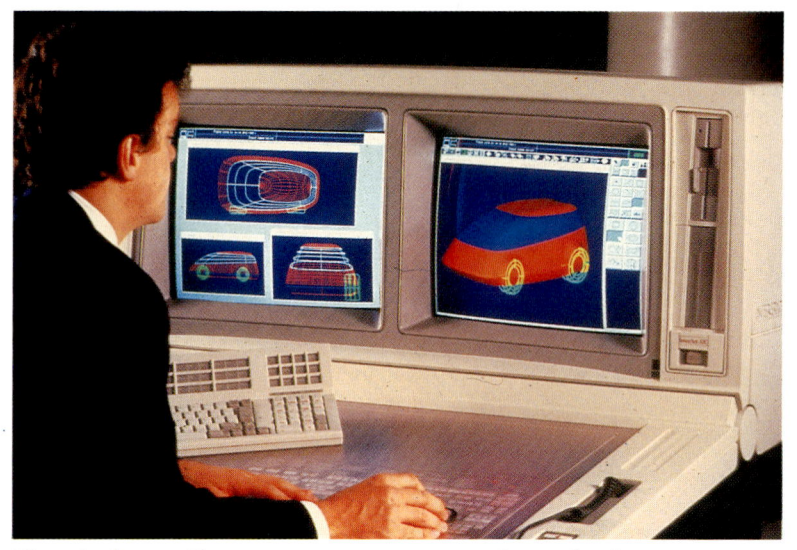

Plate 5 *above.* This two-screen system speeds up the drawing process and gives instant viewing from all sides. Modifications are easily perceived. (Reproduced by courtesy of Intergraph (UK) Ltd.) See pp. 69, 73, 74.

Plate 6 *below.* The intricacies of circuit boards are designed with a computer. Shown here is the first photographic stage in the process of manufacturing printed-circuit boards with both negative and positive images. (Reproduced by courtesy of Kodak Ltd.) See p. 74.

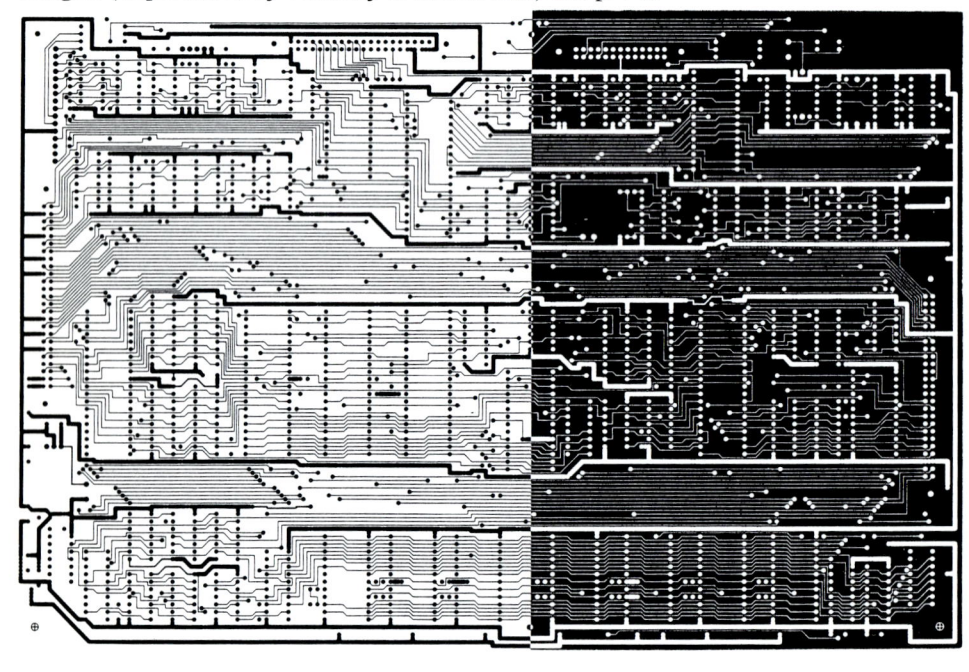

Computer-aided design in engineering

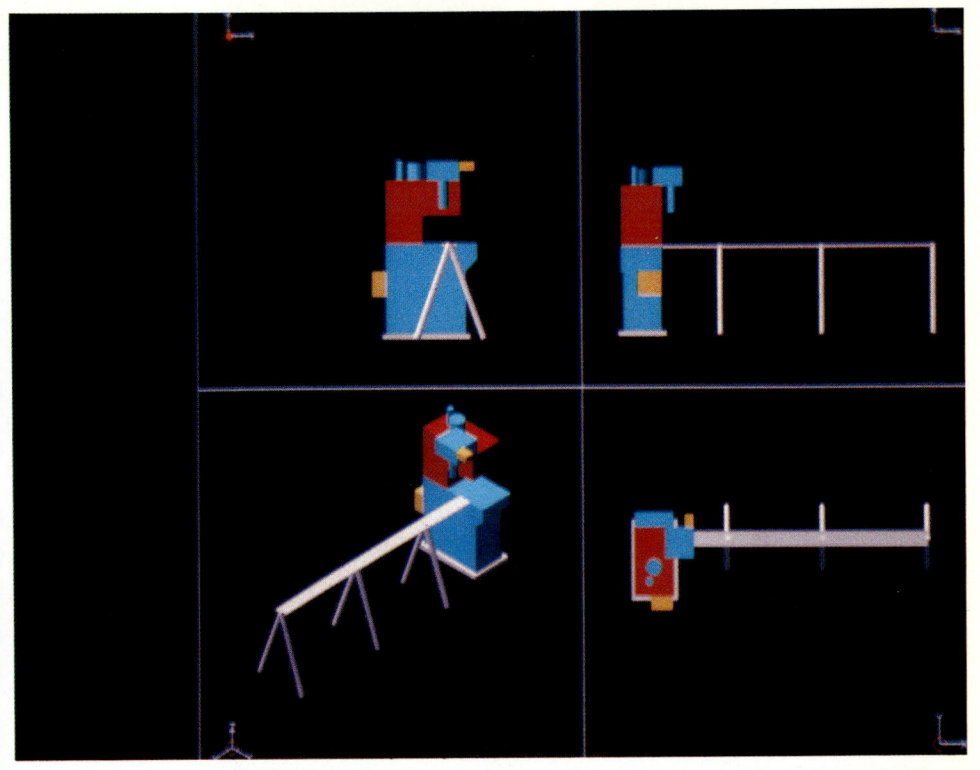

Plate 7 *above.* Using basic shapes the designer can twist and turn his design to see the combined object from any angle. (Reproduced by courtesy of J. Driscoll, University of Liverpool.) See pp. 69, 74.

Plate 8 *below.* Two stages in solid modelling techniques for component design. (Reproduced by courtesy of the University of Liverpool.) See pp. 69, 74.

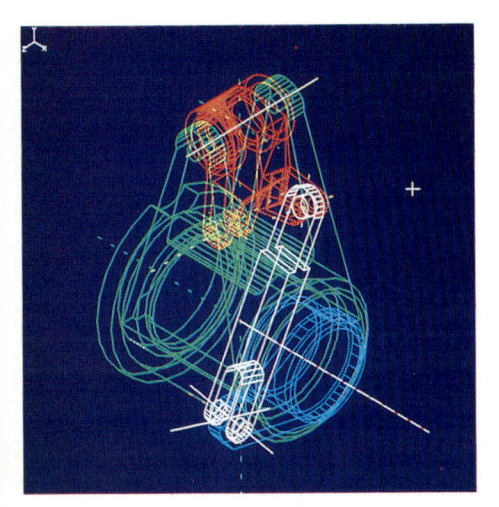

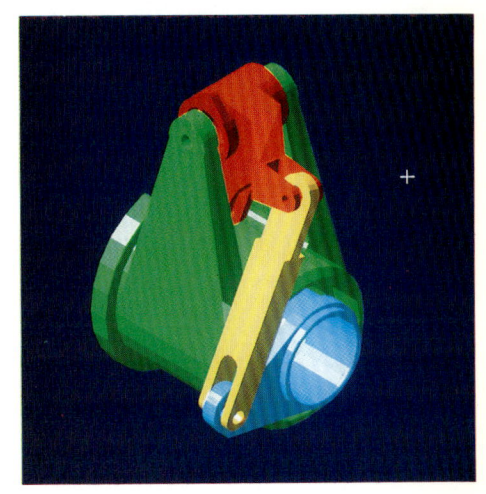

Chemists need to invoke the third dimension

Plate 9. Hand-built models old and new:
(a) *above, left*: a helical structure built of wire and cut-out pieces of plastic, about 1 foot high (1953). (Reproduced by courtesy of M. F. Perutz.)
(b) *above, right*: helical structure built of molded plastic units (1980s). (Reproduced by courtesy of the BBC Open University Production Centre.) See p. 77.

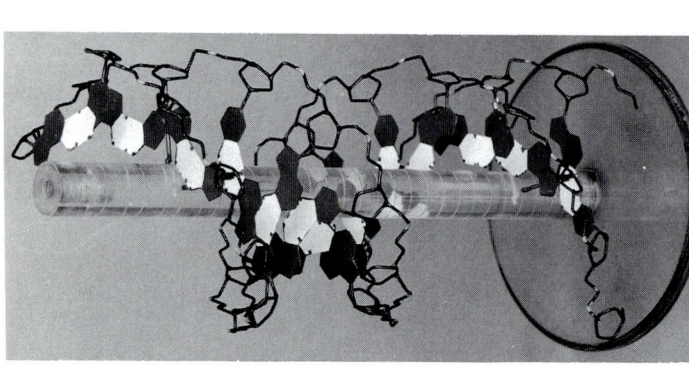

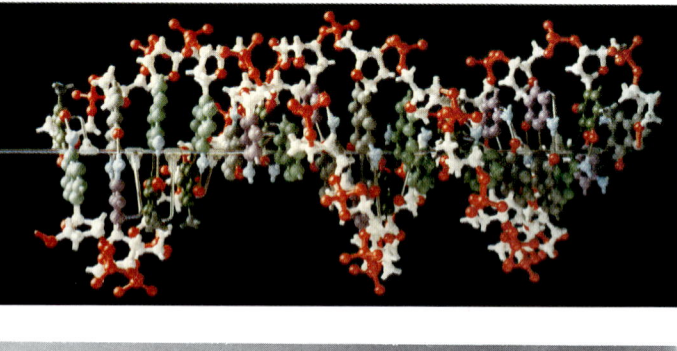

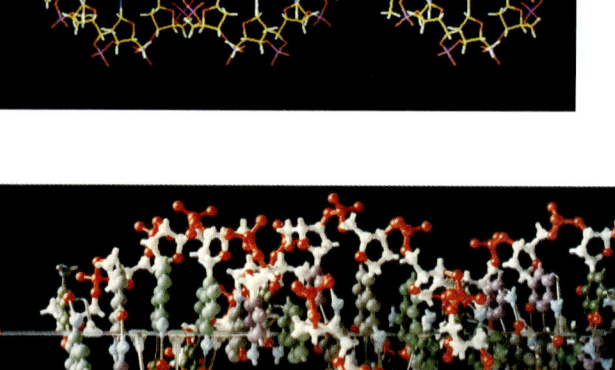

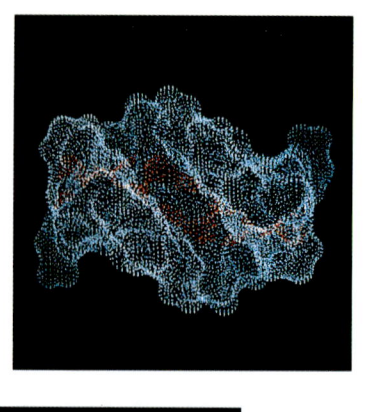

Plate 10. Computer-generated models:
(a) *above, left*: stick representation of B-DNA.
(b) *right, top*: space-filling representation of the cro protein.
(c) *right, bottom*: surface representation of B-DNA.
(Reproduced by courtesy of S. Neidle and the Cancer Research Campaign. Biomolecular Structure Unit at the Institute of Cancer Research.) See pp. 77, 143, 146.

Astronomy pictures

Plate 11 *above.* Image of the Great Red Spot of Jupiter obtained by Voyager 1 at 1.8 million km from Jupiter. (Reproduced from *Beyond vision* by courtesy of Oxford University Press.) See p. 82.

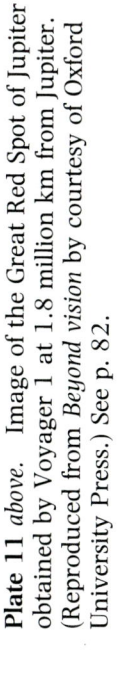

Plate 12 *above, right.* The Rings of Saturn picture obtained by Voyager 2. The colour variations indicate different chemical compositions. (Reproduced by courtesy of the Jet Propulsion Laboratory.) See p. 82.

Plate 13 *right.* A comet on collision course with the sun. This 'kamikaze comet' was discovered by a satellite in 1979. False-colour reconstruction. (Reproduced from *Beyond vision* by courtesy of Oxford University Press.) See p. 82.

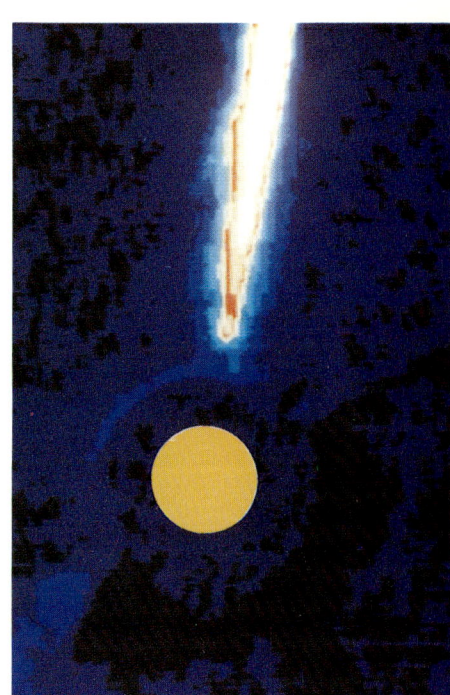

Computer-processed images for the physician

Plate 14. Recording activity in the brain using the 'invasive' positron-emission technique. The patient's head is surrounded with a 'halo' of gamma-ray detectors which feed signals to a computer. The computer builds up images of 'slices' through the brain—hence the term tomography. The patient has been injected with a positron-emitter that attaches to glucose in the blood. The glucose concentrates in areas where there is increased metabolic activity, which in turn shows up as increased amount of radioactivity. For these four scans the patient was given different kinds of auditory stimulation. With only verbal stimuli (a Sherlock Holmes story) the left side of the brain appears more active; with non-verbal stimuli (a Brandenburg concerto) there is more activity in the right-hand side. (Reproduced from *The particle explosion* by courtesy of Oxford University Press.) See pp. 96, 107.

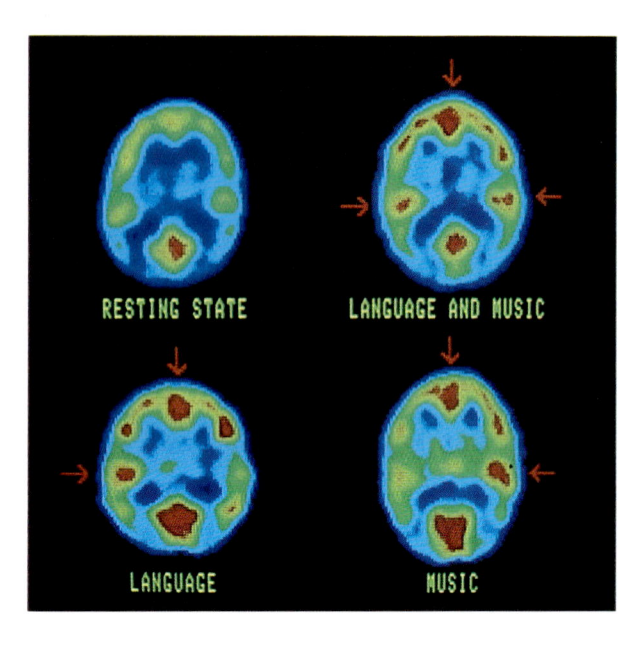

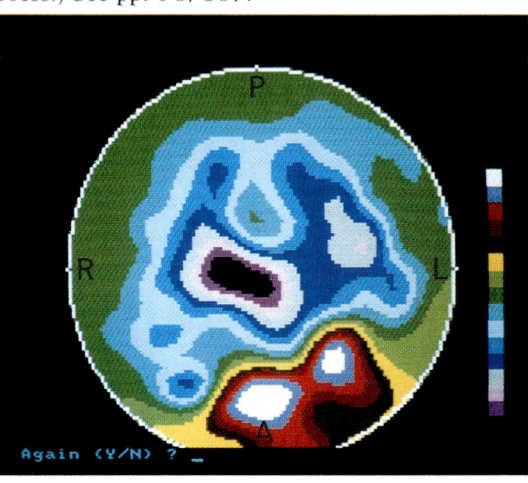

Plate 15. Image showing blood flow in the heart and lungs obtained by a non-invasive technique using the passage of low-frequency electric currents through the body from the skin surface to measure rates of blood flow. (Reproduced by courtesy of Sheffield University and Health Authority, Dept of Medical Physics and Clinical Engineering.) See p. 96.

Molecular structure

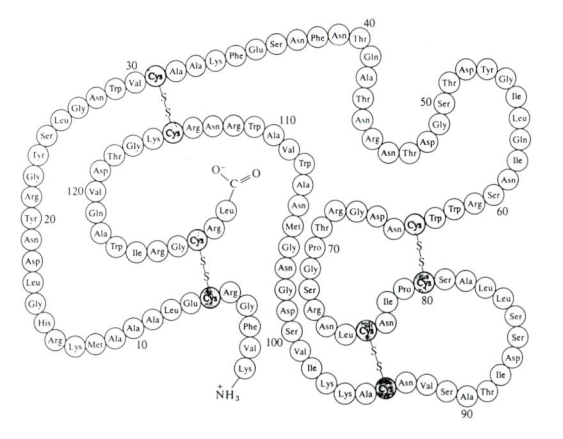

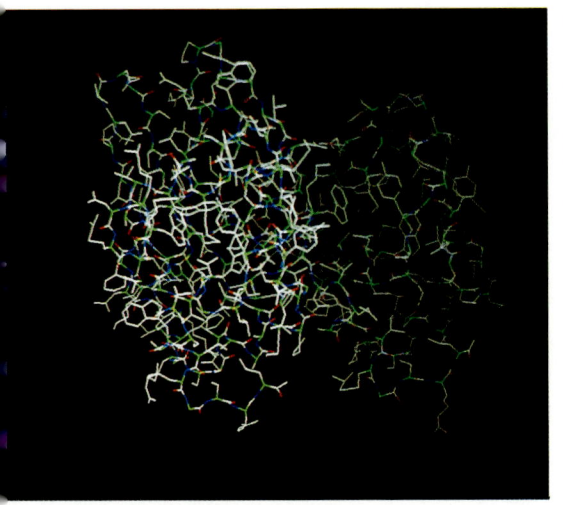

Plate 16. The giant molecule lysozyme.
(a) *(top)* Its amino-acid chain.
(b) *(below)* A stick representation of structure. (a) Reproduced from *Organic chemistry* by courtesy of Benjamin/Cummings Publishing Co. Inc. (b) Reproduced by courtesy of S. Neidle and the Cancer Research Campaign Biomolecular Structure Unit at the Institute of Cancer Research.) See pp. 99, 105, 146.

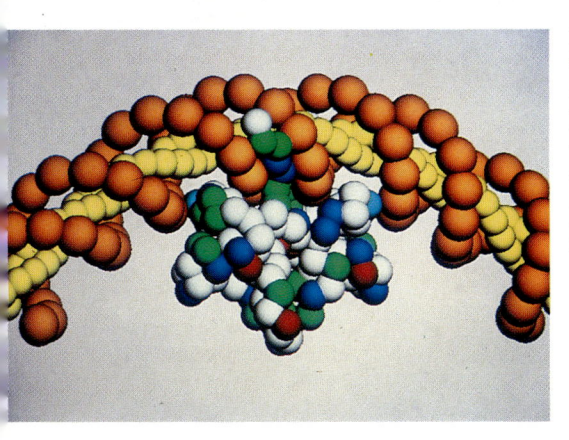

Plate 17. Receptor molecules. This diagram shows the Cro protein molecule (below) binding on to the helical DNA. Potential binding sites and 'goodness of fit' between drug and receptor molecules are subjects of research. (Reproduced by courtesy of B. W. Matthews, University of Oregon.) See pp. 99, 146.

———————— **A protein binds to a specific site on DNA** ————————

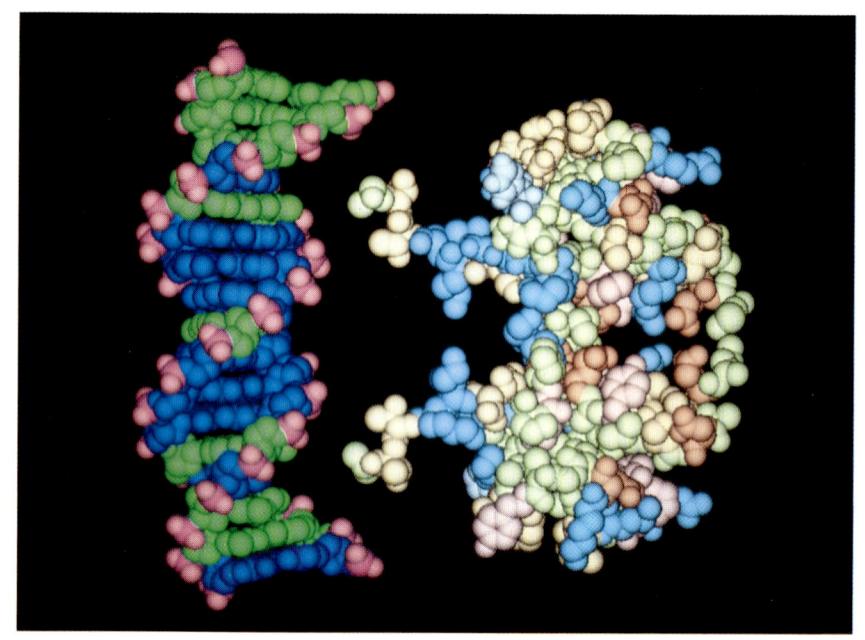

Plate 18. A computer visualization of a protein and DNA:
(a) *above:* approaching each other
(b) *below:* bound up with one another
(Reproduced by courtesy of S. Neidle and the Cancer Research Campaign, Biomolecular Structure Unit at the Institute of Cancer Research.) See pp. 105, 143.

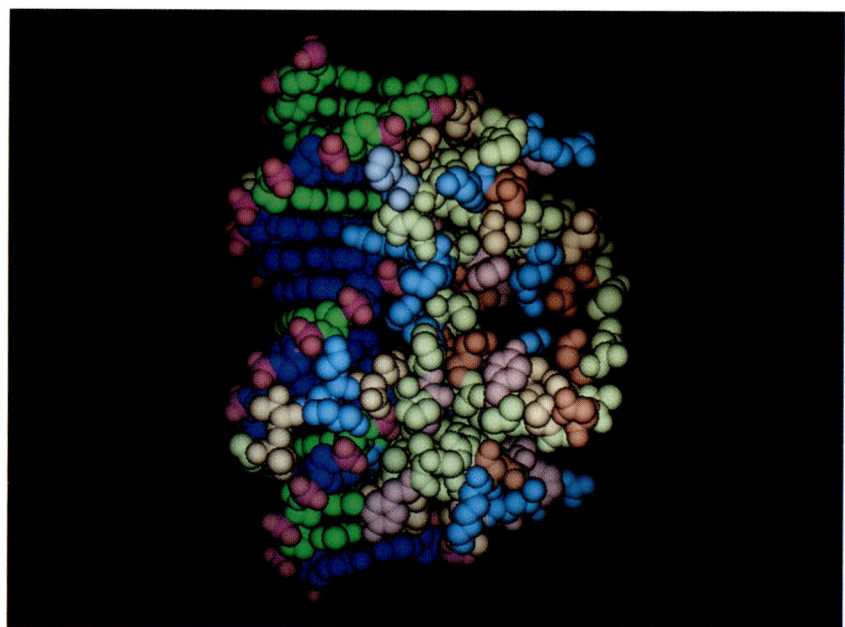

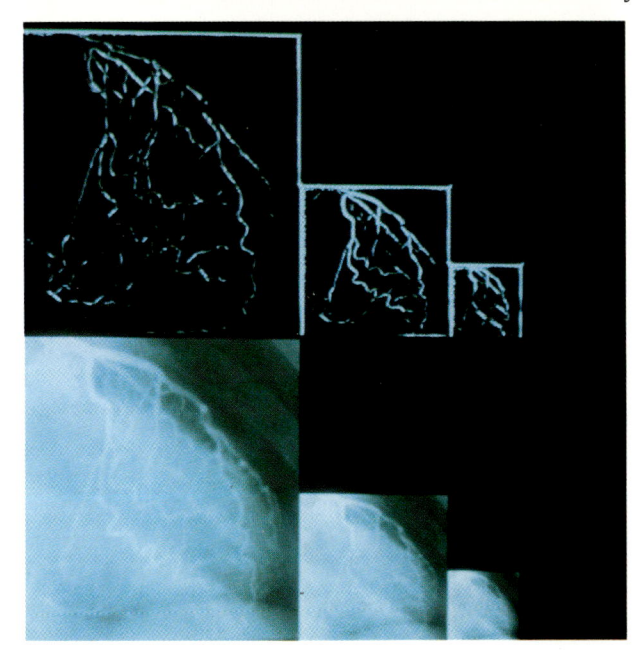

Automated detection of linear features by the computer

Plate 19.
(a) Blood vessels appear as line-like structures in a coronary angiogram. The angiogram is digitized, and lines of different width are detected at different scales. Narrow vessels are detected in the original image, broader vessels in reduced versions of the image (a). These images are then combined to form a composite one of the arterial tree as in Fig. 6.11b. (Reproduced by courtesy of the Wolfson Image Analysis Unit, University of Manchester.)

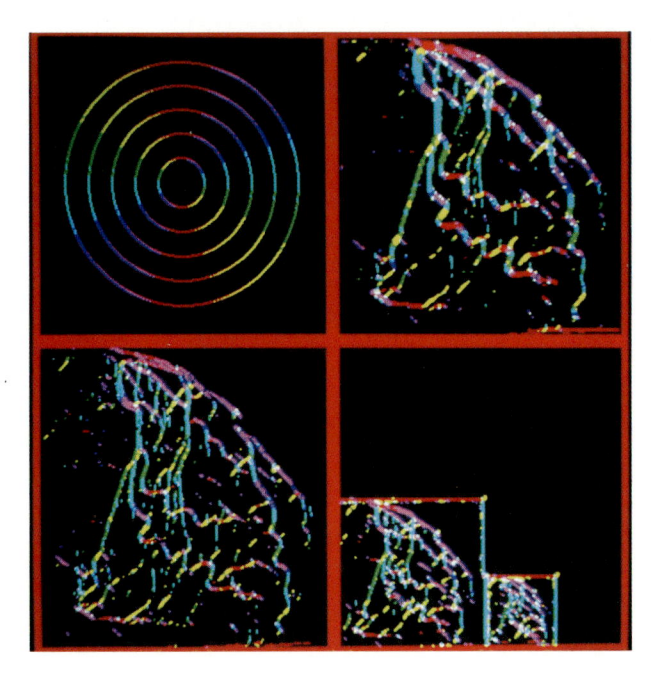

(b) Colour-coding of vessels. Horizontal lines are coded red, vertical ones blue. Top right, vessels detected at two scales are combined into one image. At bottom, vessels coresponding to three levels of hierarchial representation as above. (Reproduced by courtesy of the Wolfson Image Analysis Unit, University of Manchester.) See p. 110.

Digital mapping

Plate 20. Much of the manual work of cartography is removed by the computer. Cartographers, using a touch-pen, can make new maps, amend and up-date old ones, and enlarge and reduce maps, as well as over-lay and merge data from different sources. (The SysScan System. Reproduced by courtesy of SysScan UK Ltd.) See p. 119.

Mapping by satellite

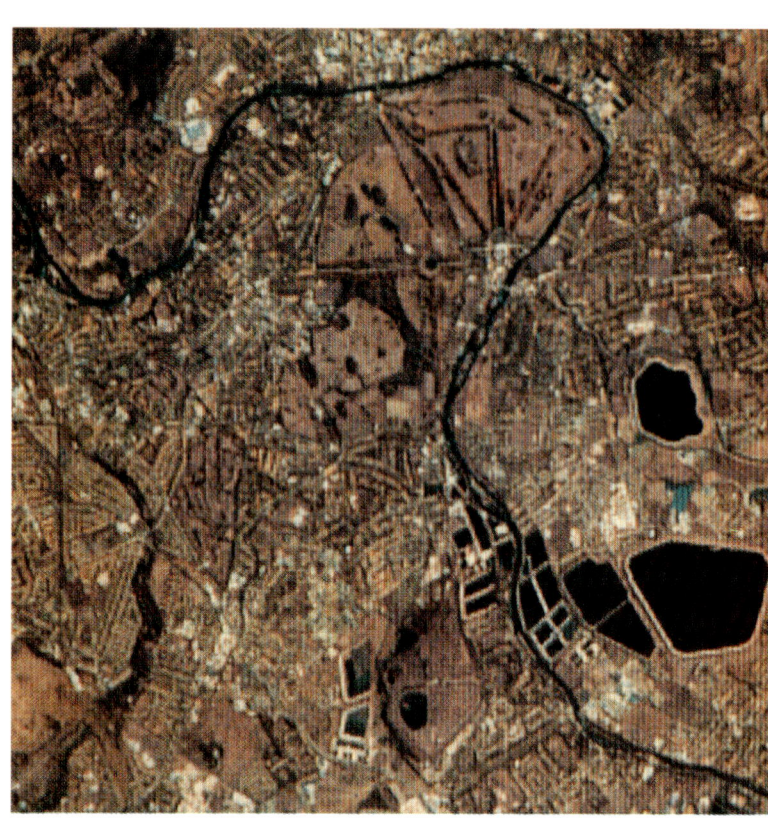

Plate 21. A satellite scan of south-west London. The colours are false. Water in reservoirs and rivers is black; Hampton Court Park in the bend of the river is brown. Scan lines can be seen at the south-west corner of the picture. (Reproduced by kind permission of Clyde Surveys Ltd., Maidenhead, SL6 8BU, England.) See pp. 121, 122.

Mapping in the infra-red

Plate 22. Thermal infra-red imaging of heat loss from an industrial site.
(a) *above, left*: The exact location of all areas of heat-loss can be identified. At (A) a steam main, 1 metre below ground
  (B) north lights in roof
  (C) well-insulated buildings appear black.
(b) *above, right*: Each colour represents a 1°C range of temperature. Digital colour is produced by assigning a colour to a range of grey levels from the data tape. Each grey level has a defined temperature; thus specific temperature bands as small as 0.2°C may be assigned to each colour. The colours in ascending temperature order are black, blue, green, cyan, red, magenta, yellow, and white. (Reproduced by kind permission of Clyde Surveys Ltd., Maidenhead, SL6 8BU, England.) See p. 123.

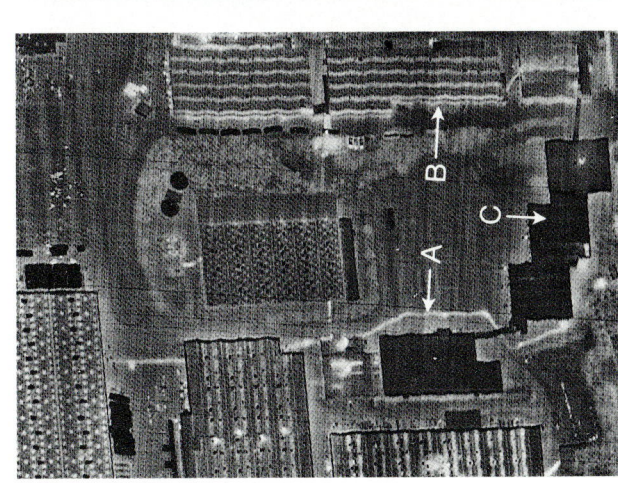

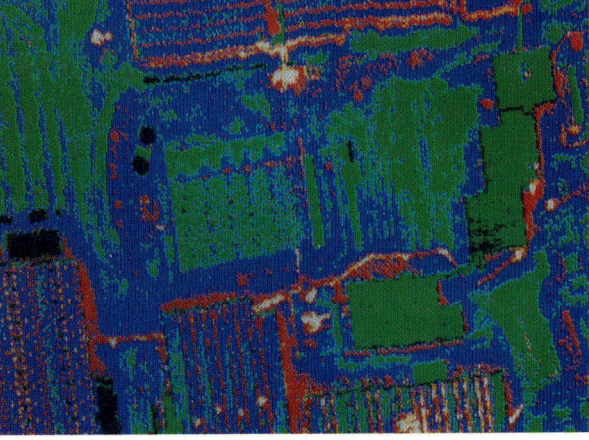

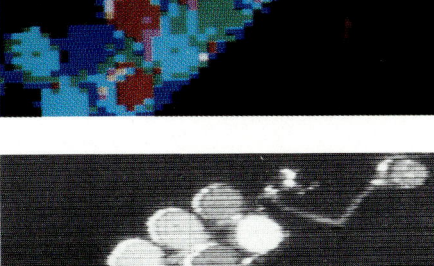

Plate 23. Oil-storage tanks showing high heat-loss.
(a) *above, left*: Magnified (enhanced) image to give greater detail.
(b) *above, right*: Colour reveals the detailed pattern of heat-loss. Each pixel on the image represents an area of 0.8 metre by 1 metre, and each colour represents a 3°C temperature step. (Reproduced by kind permission of Clyde Surveys Ltd., Maidenhead, SL6 8BU, England.) See p. 123.

Mapping in the infra-red

Plate 24. False-colour infra-red aerial photograph of part of the Great Wall of China and the semi-natural vegetation. Healthy vegetation reflects more strongly at near-infra-red wavelengths than it does at green wavelengths, and so appears in tones of red on infra-red film. Subtle changes in reflectance are related to the health of the vegetation, and also are often related to specific species. The Wall and bare soil or rock do not reflect well in near infra-red, and so appear blue and green. (Reproduced by courtesy of the National Remote Sensing Centre of China and the Natural Environment Research Council.) See p. 123.

Mapping with radio and sound waves

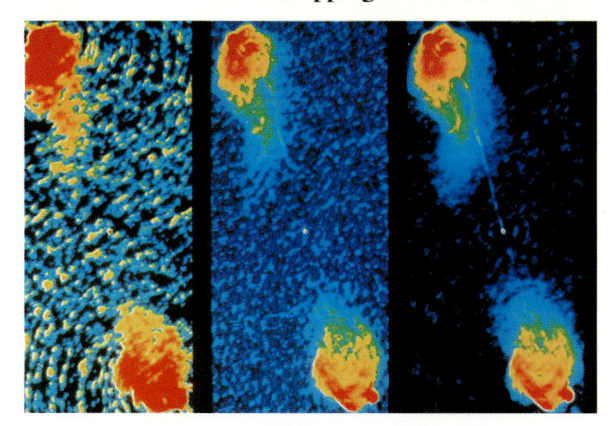

Plate 25. Radiograph of the radio-galaxy Cygnus A, showing three successive stages in the automated clearing of the radio picture. (Reproduced by courtesy of the National Radio Astronomy Observatory.) See p. 124.

Plate 26. Map, obtained by radar, of part of the eastern edge of the Andes and the adjacent Amazon Basin. Used for oil exploration. (Reproduced by courtesy of Hunting Technical Services Ltd.) See pp. 121, 124.

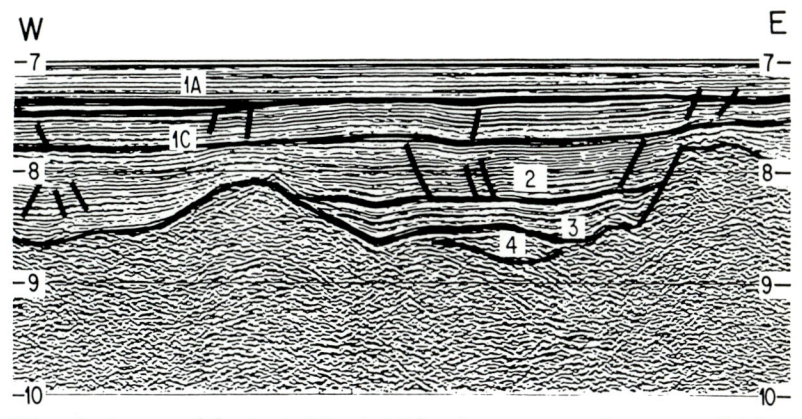

Plate 27. A picture of the invisible. A 15 km-long seismic line recorded from the sea floor off Cape Finisterre. The times taken for sound impulses transmitted from a ship to be reflected back from rock layers below the sea floor are transformed into a picture of strata 'seen' in a vertical section. (Reproduced from Rift tectonics on the passive continental margin off Galicia in *Marine and Petroleum Geology*, **Vol. 4** (1987) by permission of the publishers, Butterworth & Co. (Publishers) Ltd. ©.) See pp. 78, 82, 124.

Decoration in the home

extraordinary! Truly a world first, the Laser FX is exclusively available from Innovations. It's very portable and compact 5"x 12" comes complete with mains cord and hi-fi connections, and brings thousands of pounds worth of laser technology into your home. You've never seen anything like it – literally!

Laser FX Transducer

SW5197 Laser FX **£149.95**

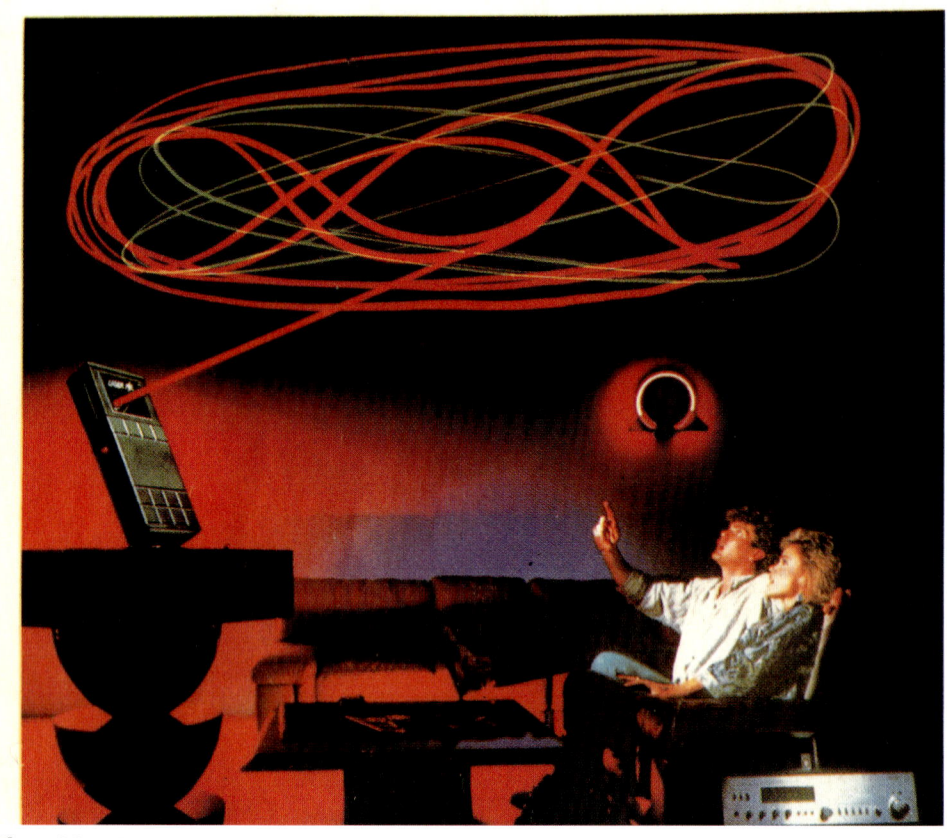

Plate 28. Item in INNOVATIONS mail order catalogue 1988. A laser transducer for use at home. (Reproduced by courtesy of Innovations (Mail Order) Ltd.) See p. 142.

_____ **Investigation** _____

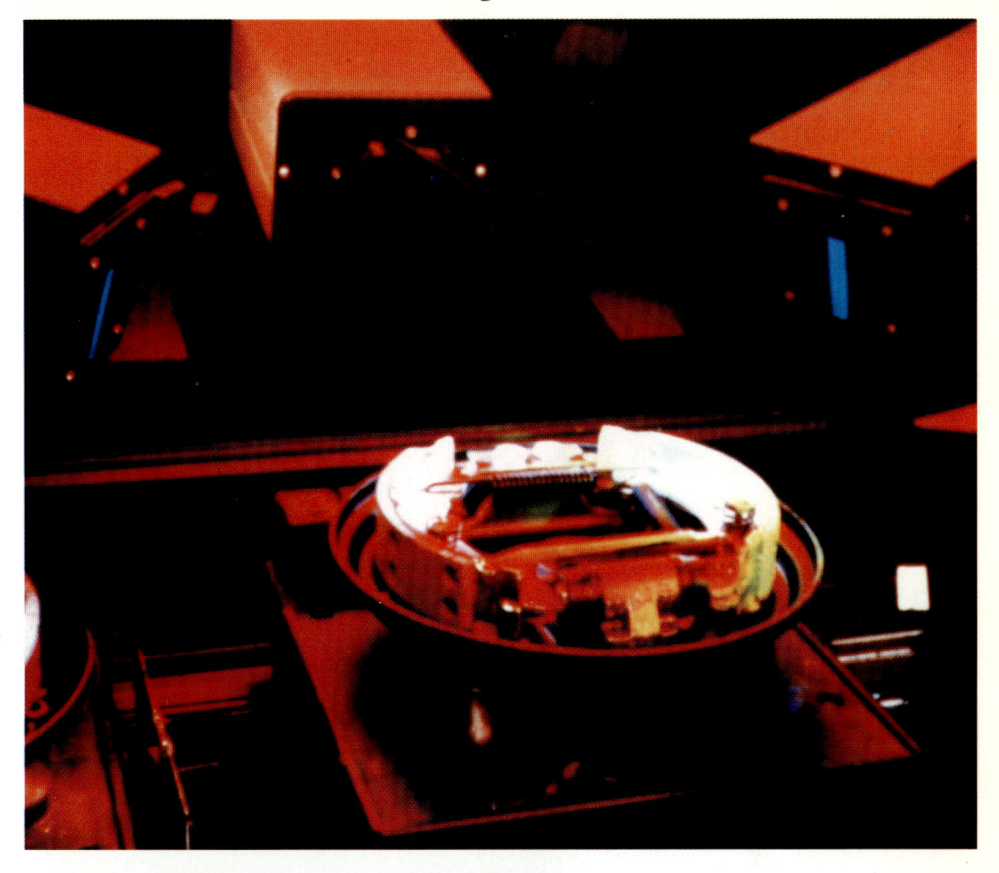

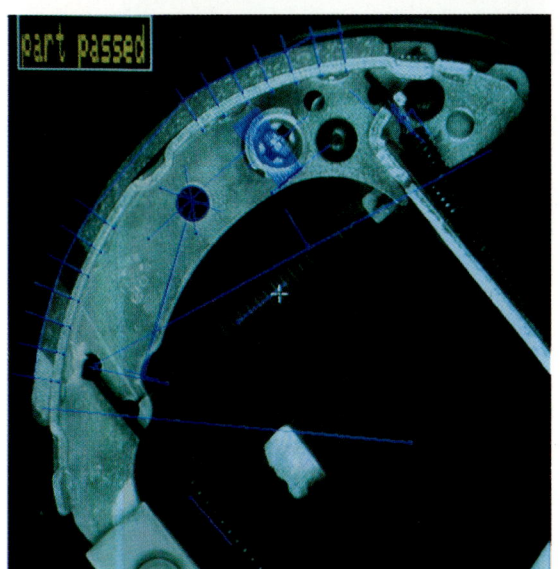

Plate 29. Brake inspection. An automatic visual system inspects brake assemblies in 5 views, makes 43 measurements in approx. 7.5 seconds, and verifies correct assembly of 36 000 brake units in one week.
(a) *above:* the inspection cell
(b) *below:* plan view of brake
   with superimposed
   geometrical model.
(Reproduced by courtesy of the Wolfson Image Analysis Unit, University of Manchester.) See p. 146.

Investigation

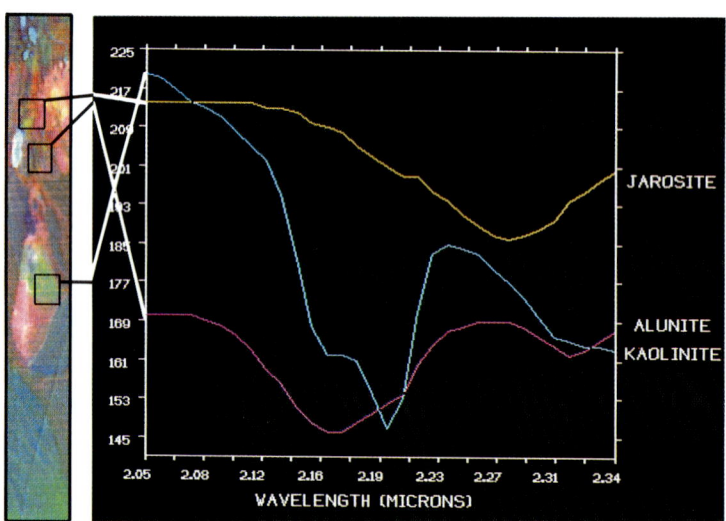

Plate 30 An image of part of Nevada obtained from a satellite. In arid regions where vegetation cover is limited, remote sensing can be employed to identify and map those minerals that exibit a unique spectral response. Sensor systems known as 'imaging spectrometers' provide the diagnostic information. (Reproduced by courtesy of Nicholas Drake and Steven Mackin.) See pp. 78, 121, 123, 146.

from the difficulties in storing, indexing, and retrieving graphic material is growing and serious, and the trouble does not stop there.

In the 1950s it became increasingly clear that the representation of chemical structures by flat ideograms that could be drawn on paper was inadequate. Ever since the 1880s, it had been recognized that the third dimension had to be invoked in order to understand chemistry. However, just as the engineers made do with projective drawing to represent three-dimensional objects, so did the chemists. The fortunate coincidence whereby many important structures were flat, or nearly so, helped with the problem. Before the arrival of computer graphics, chemists were making growing use of the clumsy method of physically modelling their structures with wire and plastic (Plate 9). These models, of course, could not be published as such, with consequent misunderstanding of the literature. Computer graphics were, therefore, as welcome to the chemists as to the engineers, and a whole new, and very expensive, skill has now grown up in the representation of protein and other structures (Plate 10). The use of any Wiswesser-type language of alphanumeric strings has been far outstripped, although the new three-dimensional knowledge can only be indexed very crudely. The verbal, alphanumeric, part of the technical literature is indexed with keywords, whose construction requires some care and dedication on the part of library professionals. The process is, however, still unsatisfactory, because of the poverty of interest shown by most chemists in the part they must play as authors—that of good abstracting and primary keywording. Consequently, in the absence of an adequate three-dimensional syntax, the literature of crucially important complex molecules is yet another Tower of Babel. There is nothing corresponding to the indexing and storage of video material by systematic alphanumeric methods, as carried out by the BBC or in sophisticated drawing offices. We have here a collection of interacting urgent needs and opportunities.

For those chemists who like evolutionary advance, the work is twofold. First, they can try to make the computer-retrieval of chemical structures more user-friendly by applying the pattern-recognition techniques developed in other fields. They could thus create better procedures than those of the literal string-translation of Wiswesser. Secondly, they can work, like Rosalind Franklin, on the science infrastructure needed for a new Crick or Watson to discover the nature of the brain's cryptology and chemical memory. It is hard to know which kind of science-building will provide the keys, the clues, and the methods. So it is good that there are plenty of dedicated researchers for the job.

Then there are at least two other tasks for the chemists who are inclined to engage in revolutionary thinking. First, it will be up to them to provide the flashes of insight and of lightning needed to see what kind of user-friendly system will grip the chemical community. Sir Derek Barton provides one exemplar: he caught the chemists' imagination by his brilliant treatment of conformational analysis, which deals with the rules governing the three-dimensional arrangements of atoms and their bonds within molecules. An initiative of this kind could get us all into computerized chemistry.

Pictures in the physical sciences: physics, astronomy, and earth sciences

Physicists, astronomers and earth scientists have had major incentives to automate both the recording of pictures and the playback of processed data (for example from satellites). Their equipment can provide Niagaras of data about stars, particle tracks (Figs. 5.7, 5.8), spectra of every kind (Fig. 5.9, Plate 30) (a matter now also of great concern to the chemists) diffraction patterns, microscope particle counts (where biologists are important customers), subsurface rock structures (Fig. 5.10, Plate 27), and many other information streams. Their own fields of study have stimulated inventive work about the processing of a variety of data. Physicists both invented the laser and interferometer and use them in data processing. The early computing that went into the automation of telescope sky surveys, such as the Edinburgh Galaxy machine (now Cosmos), was directly usable in automated microscopy as applied in the Coulter Counter for counting images in a microscope's field of vision. Physicists and astronomers were also the pioneers in the use of both pictures and numbers in the testing of their models and theories. When number processing was tedious, physicists used mathematical transformation to devise plots, preferably linear ones, where correlation could be simply tested, with allocation of error limits (Fig. 5.9). The acquisition of the meaning 'clear' by the word 'graphic' is partly a consequence of the ease of seeing a conclusion from the plotting of a well-designed graph based on a good physical model (Fig. 5.2, p. 70). The fact that the physical chemists and biologists have now adopted widely the mathematical approaches of the physicists serves to emphasize the value of the pioneering work in physics itself.

Fig. 5.7 1930 Classic photograph taken in a cloud chamber, showing alpha particle tracks (resembling a broomstick) with a proton track (leftward streak at top). (Reproduced from *Beyond vision* by courtesy of Oxford University Press.)

Fig. 5.8 1987 Computer diagram of tracks. The computer can calculate and draw in paths from mathematical information. (Reproduced from *The particle explosion* by courtesy of Oxford University Press.)

Pictures in the physical sciences: physics, astronomy, earth sciences 79

————————— **Pictorial recording of particle tracks** —————————

Fig. 5.7

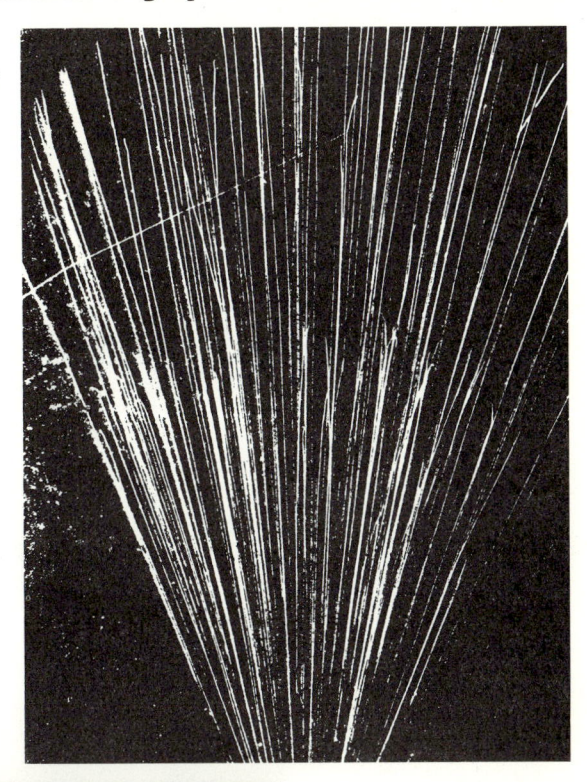

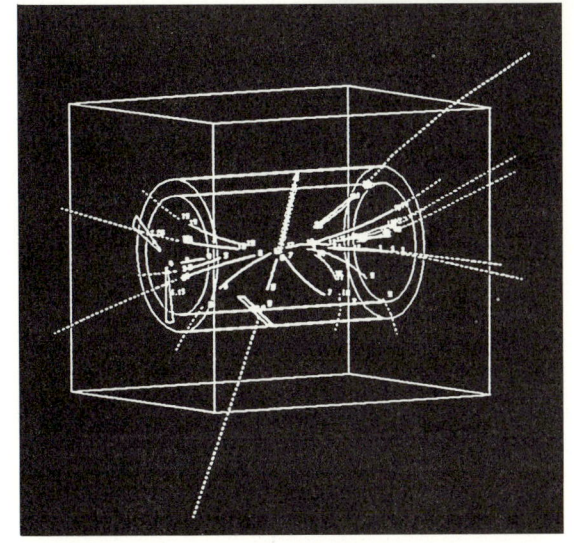

Fig. 5.8

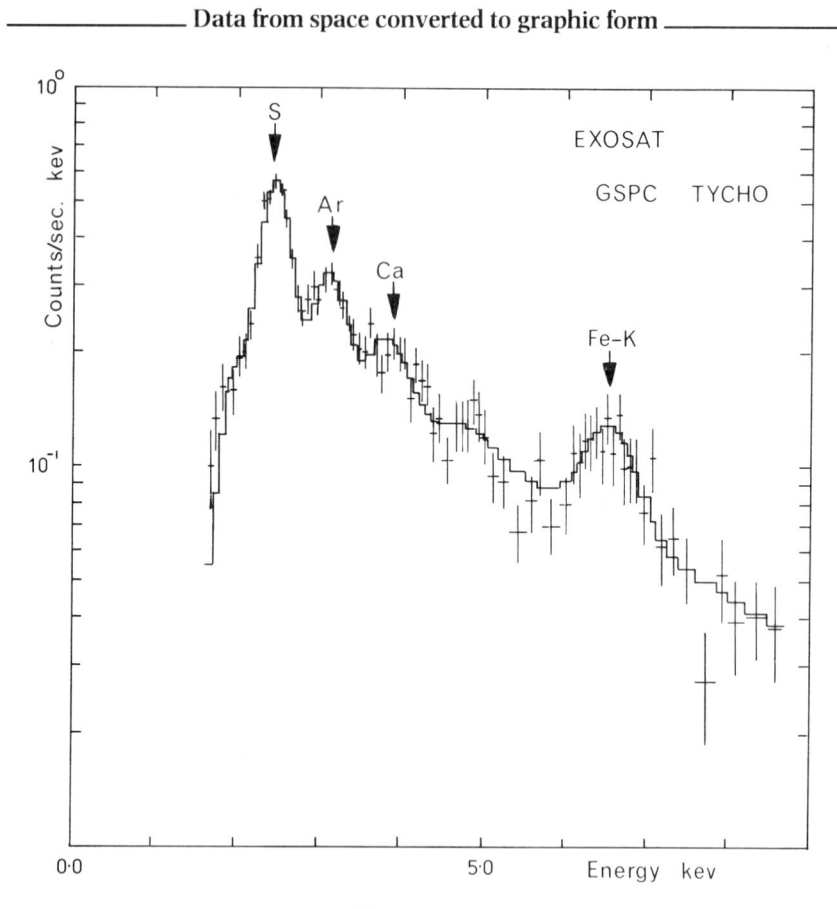

80   *Pictures and words in engineering, physics, and chemistry*

——————————— **Data from space converted to graphic form** ———————————

Fig. 5.9 X-ray spectra. Spectral lines of sulphur, argon, calcium, iron, and potassium from the remnants of Tycho's supernova of AD 1572 as observed by astronomy satellite Exosat four centuries later. (Reproduced by courtesy of M. Rowan Robinson.)

Pictures in the physical sciences: physics, astronomy, earth sciences 81

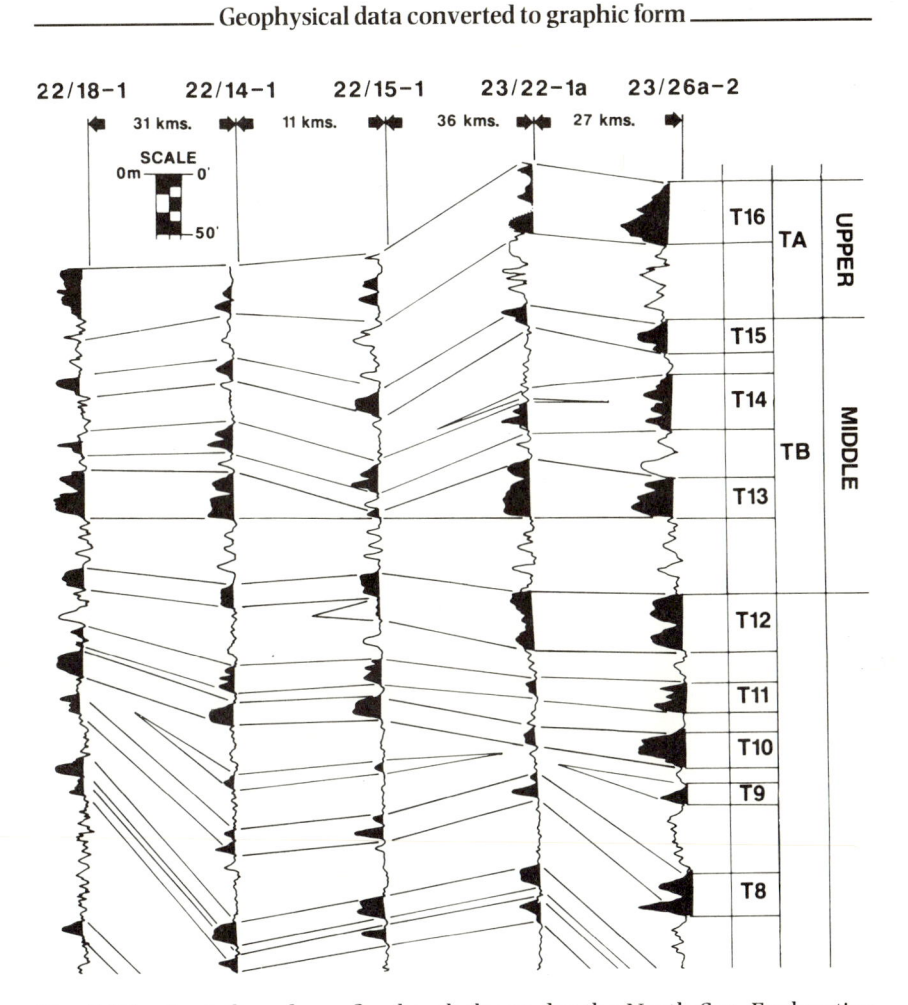

────────────── Geophysical data converted to graphic form ──────────────

Fig. 5.10 Sonic logs down five boreholes under the North Sea. Exploration geologists use sonic data to detect porosity that may contain oil or gas, and to correlate strata from one borehole to another (for example, stratum T.15). (Reproduced from 'Allochthomous chalk group members' in *Marine and Petroleum Geology*, **vol. 3** (1986) by permission of the publishers, Butterworth & Co (Publishers) Ltd. ©.)

One of the main applications of more sophisticated pictorial displays in physics, astronomy, and earth sciences arises from the desirability of converting masses of observational data into a pictorial form (Figs. 5.9, 5.10). Such displays can indicate more clearly what is happening, in advance of clarification by graphical and algebraic methods. The growing popularity of astronomy on television is providing the incentive for indicative colour presentation, along the lines set up in the encounters of space probes with planets and comets. Millions of television viewers saw the tail of gas and dust behind Halley's Comet (recorded by the spacecraft Giotto) with its concentric bands of computer-generated false colour. There is still a lot of confusion about the meaning of the different colours, but the interest of a wider public has been attracted, which is the standard prerequisite for education (Plates 11, 12, 13).

The applications of pictorial methods to the resolution of problems in physics are legion, and are employed in, for example, geophysics and geology (Figs. 5.10, Plate 27), meteorology (Figs. 4.4, 4.6, 7.7), molecular physics (Figs. 5.11, 5.12), solid-state and grain-boundary phenomena, astrophysics and cosmology (Fig. 5.9), electromagnetic spectroscopy, mass spectroscopy, magnetic resonance phenomena, and in the new science of chaos (Appendix B), where, as Gleick puts it in his book, 'it's masochism for a mathematician to do without pictures'.

Why video reports might be better than written ones

Much of the published literature that deals with science and engineering is not as good as it should be, from the viewpoint either of clarity or effectiveness. The part that is aimed at professional readers varies from the excellent to the atrocious. That which is written for the general reader is incomparably better

---

Fig. 5.11 Hand-built atom-scale model ($\times 15$ million) of a structure of the kind that can be manufactured by molecular-beam epitaxy. The electronic properties of the dark and light layers are different, and so they can be exploited in devices such as lasers, optical switches, and ultimately optical computers. (Reproduced by courtesy of the University of Liverpool.)

Fig. 5.12 Model of a (111) surface of silicon showing the atom positions which occur as a result of reconstruction of the atomic bonds to minimize surface energy. The white outline identifies the unit cell of the surface. The model is based on images, such as the inset picture, obtained using a scanning tunnelling microscope, which measures the potential associated with individual atoms on surfaces. (Reproduced by courtesy of IBM UK Ltd.)

Why video reports might be better than written ones 83

———————————— Modelling in physics ————————————

Fig. 5.11

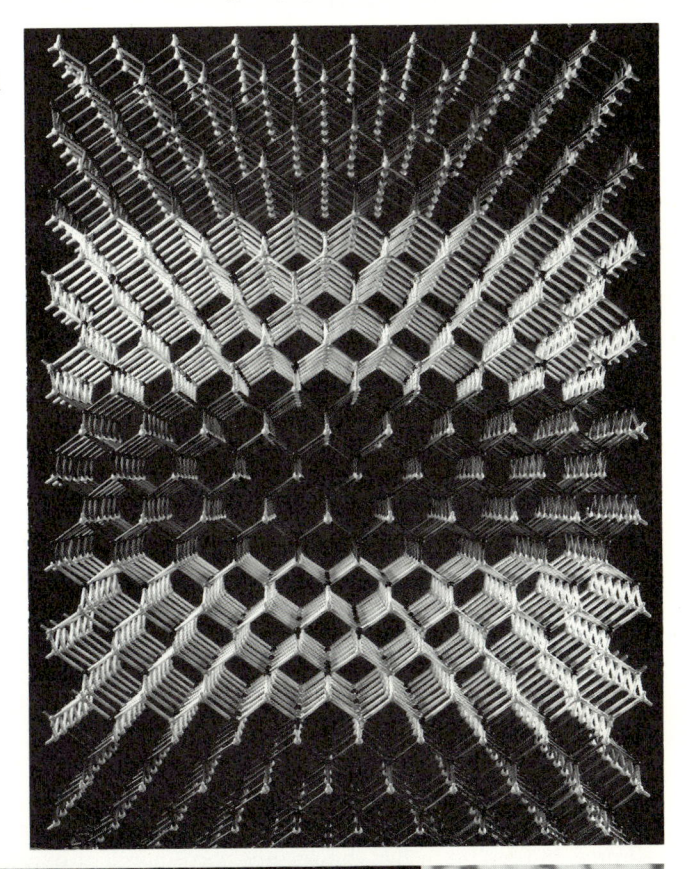

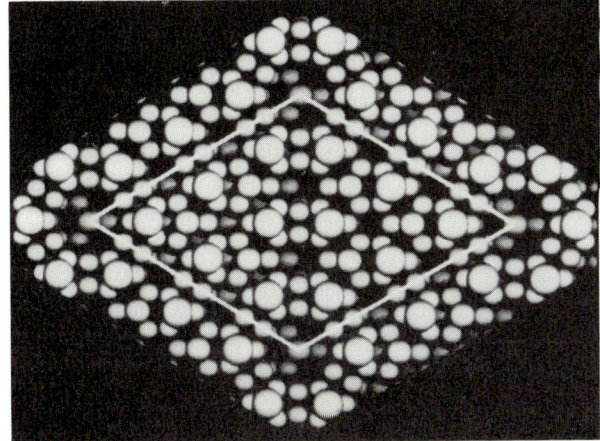

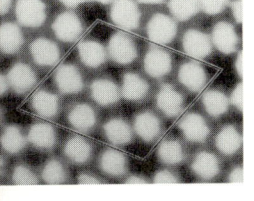

Fig. 5.12

than it was twenty years ago, but could still be improved. It is the private literature—internal reports within companies or reports written for funding agencies—that is worse than any. (Chairmen of the Divisions of big companies have been known to muddle their reports deliberately so as to prevent rational allocation of capital and leave it to the lobbying process, which may offer better prospects.) Nor is the patent literature the clear disclosure of technical art that the law intended.

The situation is what might be expected from the nature of the task and of its (usually excellent and dedicated) practitioners. In most other fields of writing, the authors took on their jobs partly because these involved writing, which they wanted to do. Advertising executives, reporters, playwrights, novelists, and editors chose their jobs because they wanted to write. If their work is not clear and convincing to the customer, they suffer. Lawyers do not write. They edit or adapt orthodox forms of words to describe familiar situations, such as wills, conveyances, agreements; the criteria are clear: lack of ambiguity, conformity with statute law and precedent, and accurate representation of intention. Here, readability and communication of ideas to a lay audience are not of the essence. Accountants and bankers write virtually nothing: managers have others to write for them. Scientists and engineers and teachers in higher (but not in school) education generally, on the other hand, are expected to write, and do so in quantity, as a by-product of their main work, which is to find things out or create products or processes.

The volume of scientific and engineering writing in relation to the number of operators is large. Yet most of those involved chose their jobs with no thought of the writing that they would be called upon to do. Indeed, knowledge of the lifetime demands of writing might have put some of them off.

The science journalist is probably the best placed. At least he chose to write. The standard of writing in the *New Scientist* and the *Scientific American*, for example, is very high. But the science correspondent of a newspaper is less happily placed. He has an editor who wants to sell newspapers, which will not be achieved by the cautious modesty of science ('Small earthquake in Chile: not many dead'). It is hard for the science correspondent of a newspaper to compete with the paradigmatic 'Sex-change bishop in mercy dash to Palace'. He has a difficult task in attracting attention, and, when he does so, his scientific friends or peers will probably sneer; but at least the outright hostility of the scientific establishment to clear and popular writing is not what it was.

What of pictures? In view of the increased importance of pictures noted in the preceding chapters, to what extent could they be used more fully to lighten the author's task of exposition? First, economic considerations cause editors to limit illustrations and graphs to a minimum in the professional journals, although there is an increasing need for colour to attract readers in the commercial sector. Secondly, it is taking some time for scientists and engineers to appreciate the considerable power of video- and computer-graphics as a legitimate means of publication and popularization. There is also a belief that the use of these media is more complicated and more expensive than is, in fact, the case. But, thirdly, the very factor which, in the end, could be the making of pictures—namely, the high quality of the television broadcasts that set the standards—may, at this stage, be a little daunting. How is an amateur to stand comparison with programmes like the BBC's 'Horizon'?

The answer is that he should not worry. Slickness, on the pattern of television advertising, is not what is called for. Clarity is well within the reach of anyone who is prepared to spend three days or so in learning the rules of camera work, editing, and production. Experiments on these lines have been carried out successfully. An authoritative view is that the essential message is typically contained, very economically, in a few sentences on the audio track, and that these can be spoken by a competent professional. The experimenter in person can be filmed doing something with which he is comfortable, by way of authentication, and the whole medium is still free from the orthodox inhibitions associated with the published word—such as the unhelpful referee. Examples of low-cost presentations of important conclusions or methodology are growing, and the skill needed for making animated graphics is decreasing. The commercial science journals have begun to sell video-tapes.

In this field there will also be, once again, the problem of sorting and indexing to face. Tape producers will repeatedly want to indicate the background by quoting from previous tapes of previous work. The location of these is going to be important.

6. Pictures and words in biology and medicine

Communication about and within biological systems

The subjects that could be covered in this chapter are very wide and diverse. Human records and communication in medicine and biology are strongly dependent on pictures, and have been significantly improved by recent developments. This aspect alone could provide a complete chapter. Then, perception and communication by other species is worth considerable thought. Finally, communication and recording within individual members of a species are sophisticated topics, embracing the processes whereby the species is reproduced and evolutionary change is made possible. These matters deserve our attention.

It is because of the interconnection between these topics that an attempt will be made to deal briefly with all of them in one chapter. If the book were primarily about communication, or biology, this approach would be wrong. But in order to focus on the balance between different methods of recording and transmission, it is reasonable.

Fig. 6.1 7000 BC. Aboriginal rock painting of the giant perch, Barramundi, from Arnhem Land Plateau, Northern Territory, Australia. This is an example of the X-ray descriptive style of painting, which shows internal organs and bone structure as well as the external features. (Reproduced by courtesy of George Chaloupka.)

Fig. 6.2 1633. Gerard's *Herbal*. Drawing of *Water Lillie* and *Frogge-bit*. (Reproduced by courtesy of the Trustees of the Science Museum [London].)

Fig. 6.3 1826 Part of Audubon's painting *Robin perched on a mossy stone*. (Reproduced by courtesy of the University of Liverpool Art Gallery.)

Communication about and within biological systems 87

——————— **Artists and biological illustration** ———————

Fig. 6.1

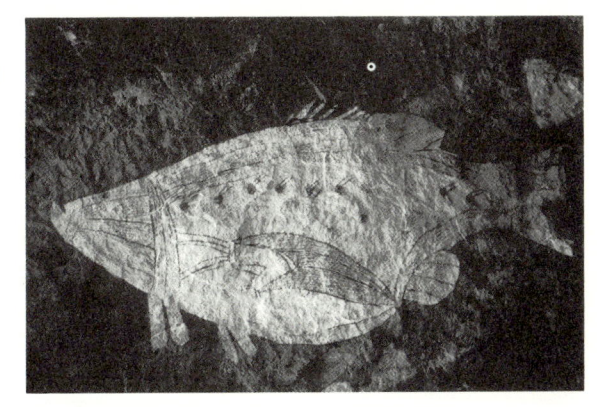

Fig. 6.2

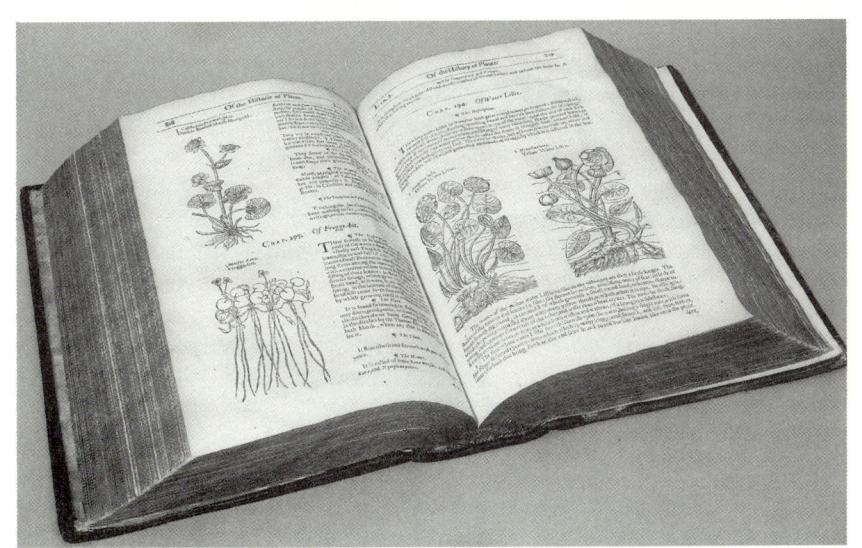

Fig. 6.3

The role of artists and photography in the recording of natural phenomena

Teaching and research in biology and medicine have always depended on the use of pictures. First-rank artists have been among the doctor's closest and most valued collaborators, and both have needed accurate knowledge of anatomy for their respective purposes. The doctors have been lucky to command such skill, including that of Leonardo da Vinci. He and others were knowledgeable about human muscles and how they change shape with movement. Stubbs pioneered the study of the horse simply in order to paint pictures of horses in action. Before his work, horses were known about only rather superficially: he actually conducted dissection of cadavers (Fig. 6.4). Moreover, artists want to convey motion, and are not content with static images. Such complexities are well beyond the descriptive capabilities of words or numbers.

The very earliest artists painted animals on the walls of caves, and thus have left us a biological record which would not otherwise have been available (Fig. 6.1). But it has been in the last three hundred years—the period of systematic science and Linnaean taxonomy in biology—that we have developed pictures for the faithful delineation of the detailed appearance of species and variants (Fig. 6.2). Amateurs have played an important part. Ornithology has been built up more by gentlemen than by players (Fig. 6.3). In particular, the invention of the microscope carried a need to provide pictures for those without one. Good photography importantly supplemented, but did not supplant, such biological and medical illustration (Figs. 6.4, 6.5, 6.6, 6.7).

Fig. 6.4 1766. Revelation by dissection. Plate XIII of Stubbs's *Anatomy of the horse*. Drawing of a cadaver. Parts were numbered and named. (Reproduced by courtesy of Liverpool City Libraries.)

Fig. 6.5 1840. Photomicrograph of insect wings (Fox Talbot). (Reproduced by courtesy of the Trustees of the Science Museum [London].)

Fig. 6.6 1988. Scanning electron photomicrograph of a coccolith in clay-rich chalk. Scale-bar 10 μm. (Reproduced by courtesy of P. Ditchfield and C. J. Veltkamp.)

Fig. 6.7 1986. Scanning electron photomicrograph of a mite on a primula leaf showing leaf hairs and stomata. Scale-bar 10 μm. (Reproduced by courtesy of C. J. Veltkamp.)

The role of artists and photography in recording natural phenomena   89

——————— Revelation by dissection, microscopy, and photography ———————

Fig. 6.4

Fig. 6.5

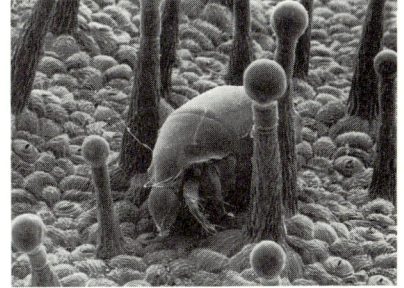

Fig. 6.7

Fig. 6.6

90   *Pictures and words in biology and medicine*

──────────────── **Foreknowledge for surgery** ────────────────

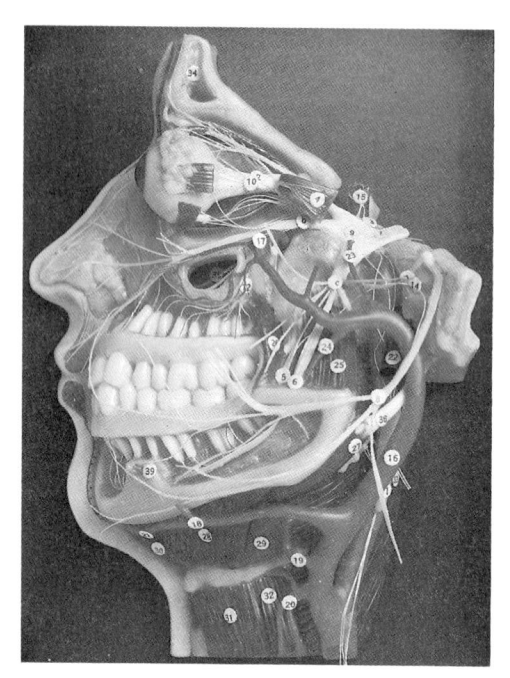

Fig. 6.8 A Victorian anatomical model of the face gave limited foreknowledge. (Reproduced by courtesy of the University of Liverpool Dental Museum.)

Fig. 6.9 Magnetic resonance tomographic image of the head. (Reproduced by courtesy of the Magnetic Resonance Centre, University of Liverpool.)

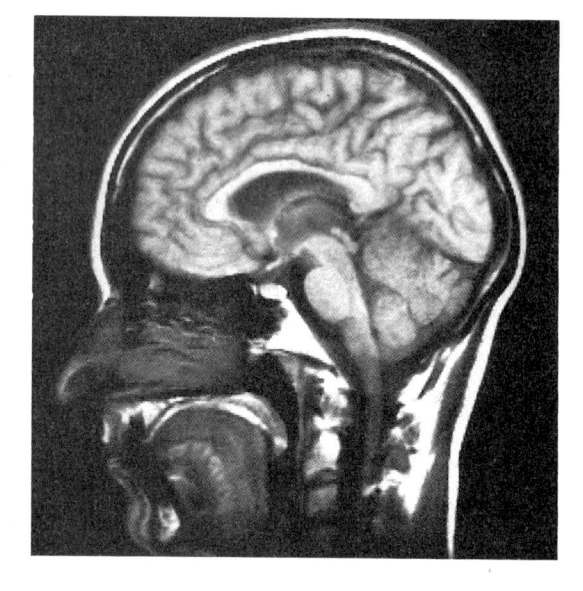

Surgical teaching and practice 91

───────────── **Foreknowledge for surgery** ─────────────

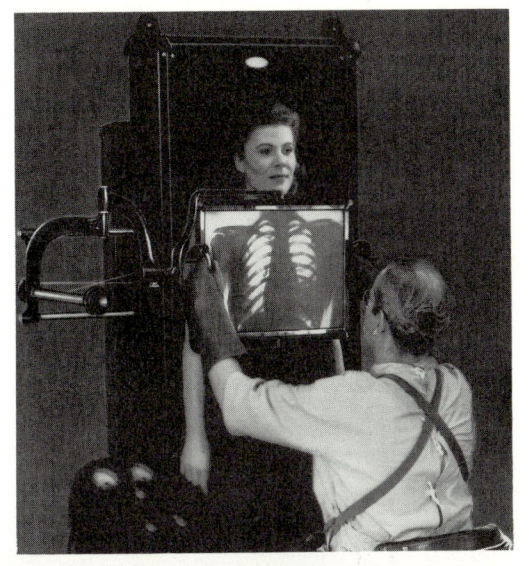

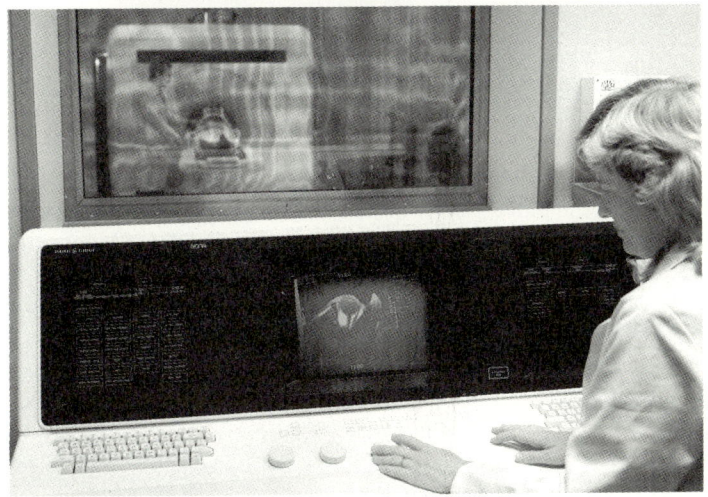

(a)

(b)

Fig. 6.10
(a) A vertical Roentgenoscope—a forerunner of today's diagnostic technology—which provided a view on a fluoroscopic screen. (Reproduced by courtesy of General Electric.)
(b) Radiographer conducting a scan with a magnetic resonance system. Image of the eye on screen. (Reproduced by courtesy of General Electric and the University of Liverpool.)

Colour has helped greatly. Yet the skilled medical or microscope illustrator still plays a part. He is able to exercise some editorial skill in emphasizing what his experience suggests may be important. The eye, backed by the brain, is a sensor of considerable sophistication, although it can increasingly be matched artifically. Pictures were filed, indexed, and sorted according to three-heading (Latin) alphanumeric names that made the data comprehensible to those with anatomical or Linnaean knowledge, although not easily to others. Thus biologists and doctors, like chemists, were the users of organized pictorial databanks well before most others, and they offer us the basis for various other beginnings.

Surgical teaching and practice

This is an area where modern pictorial technology has paid handsome dividends. The master–pupil relationship has been, and still is, crucial. No more than one or two at a time, standing by the operating table, can see what a surgeon does in enough detail to go and do it themselves. They alone can feel the sharpness and the drama of the on-the-spot decisions that depend on what is found. However, there is a lowering of tension now as new non-intrusive tomographic methods (from the Greek word for a section) give more foreknowledge (Figs. 6.8, 6.9, 6.10). These computer-assisted techniques of

---

Fig. 6.11 Coronary cineangiography is a technique whereby a sequence of X-ray images of the patient's coronary arterial system is obtained, during and after the injection of a radio-opaque contrast material directly into the vessels of interest.
(a) single frame of X-ray cinefilm. The left coronary artery and its branches are opacified (contain contrast material).
(b) a detailed map of the artery is detected automatically by the computer from image a. using new techniques (for details of method see Plate 19a and b). (Reproduced by courtesy of the Wolfson Image Analysis Unit, University of Manchester.)

Fig. 6.12
(a) A Moire Fringe Pattern generated on a normal person's back. The centres of the fringes are contours of equal height lying on the intersection between the back surface and equally spaced planes parallel to the Moire screen.
(b) The surface of the same person's back interpolated from the Moire pattern. Clinically significant parameters can be determined automatically from this recording. (Reproduced by courtesy of P. H. Dangerfield and D. Groves.)

Surgical teaching and practice 93

Computer-aided foreknowledge

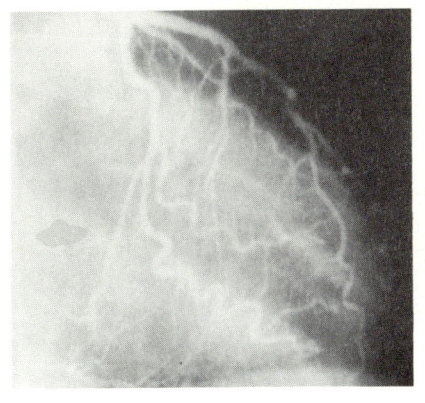

Fig. 6.11(a)

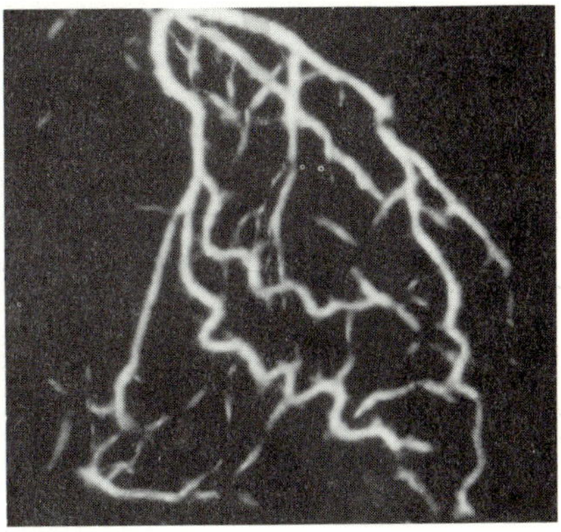

Fig. 6.11(b)

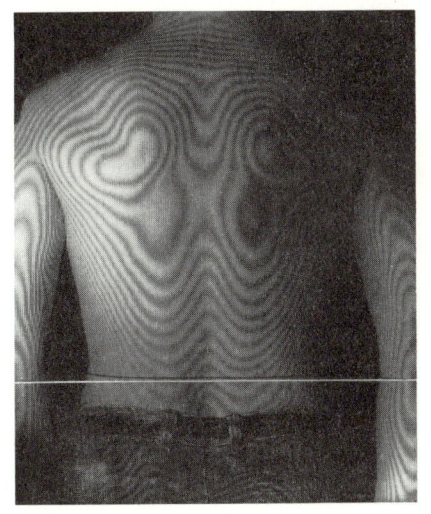

Fig. 6.12(a)

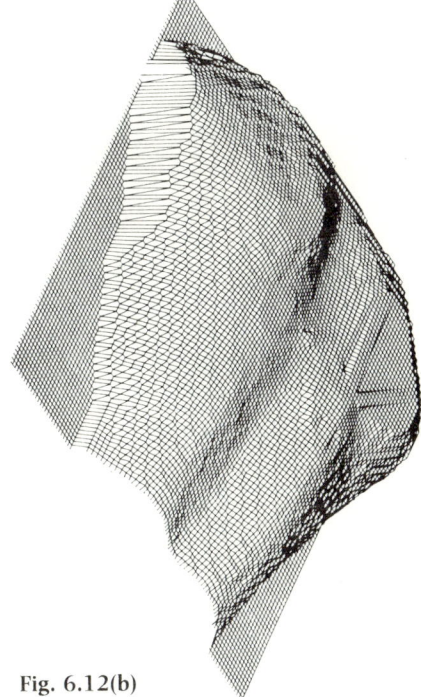

Fig. 6.12(b)

94 *Pictures and words in biology and medicine*

Computer-aided foreknowledge

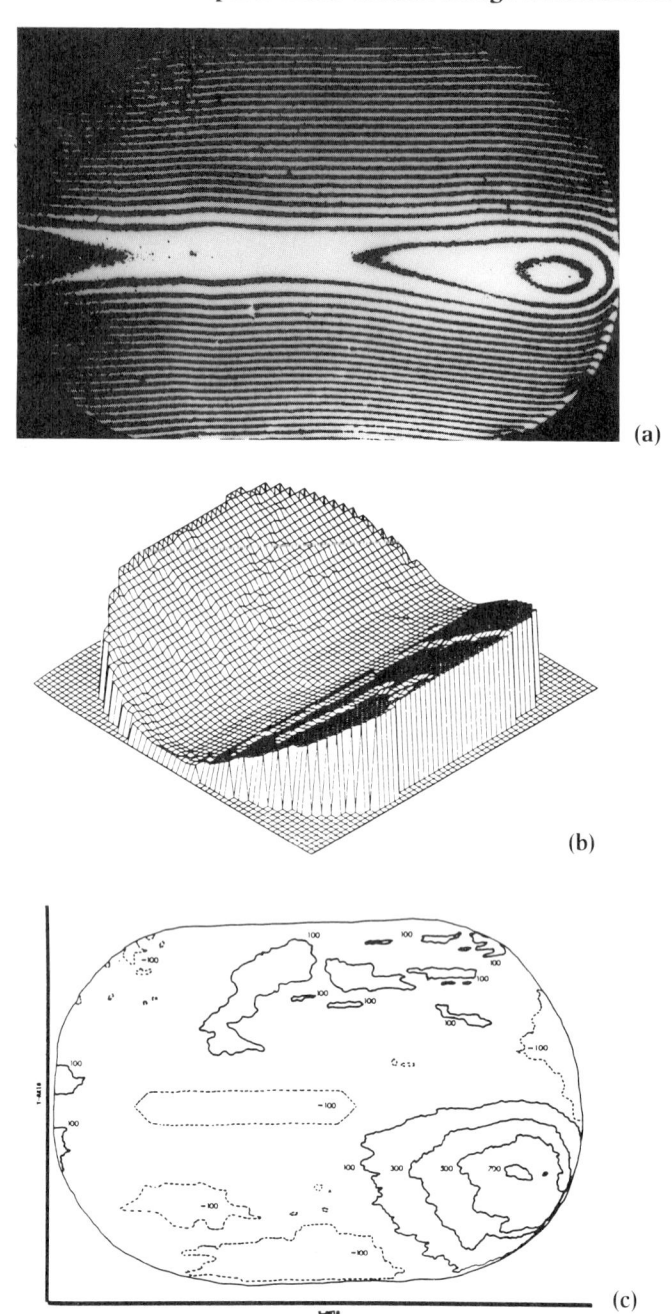

(a)

(b)

(c)

examination, whether by X-ray or magnetic resonance, confine the picture to a selected plane, so that shadows of objects in front of or behind the plane are excluded. The effect is as if a section or a slice through an organ were being examined. Other computer-aided techniques are shown in Figs. 6.11, 6.12, and 6.13. Nevertheless, dynamic images remain of the essence, and the progress of an operation can now be seen by any number of observers through closed-circuit television. The action replay, with stopped-frame procedure and time-lapse rearrangement, puts the surgeon, his successes, and his mistakes, under a very bright and searching spotlight.

With video-recording, it may be possible to look back through an operation and see if anything was left inside. This raises very difficult questions about what the lawyer or the judge should see. In a litigious country heroic, adventurous, or experimental surgery, which may be in the interests of patients, could be stopped dead. So, once again, we have an instance of a new possible need for the prevention or corruption of communication. Where ignorance is bliss, it may be folly to be wise. Who, however, draws the line?

Many examples of the utility of the televisual recording and study of surgical practice can be given, but the most spectacular is probably in the demonstration of team-surgery technology, and in other situations where the theatre environment is a key part of the new skill. Special asepsis, for example in hip surgery, is one case in point. Transplants, the use of lasers, and heart surgery involving elaborately organized auxiliary equipment and bypasses, have spread several times faster than they could have done without television and video. We do not know whether a surgeon has ever had a commentator in a different continent kibitzing by satellite, and giving advice at key points, but the idea is not absurd. It has been done many times by radio.

Physic, pathology, and the recording of correct and incorrect function

Physic and pathology do not have quite the dependence on image and dimensional precision that is with us in surgery. Verbal description, though often somewhat imprecise and lame, is certainly more useful. However, use

---

Fig. 6.13
(a) Optically contoured surface of a worn knee joint.
(b) A low-resolution example of a computer-generated surface of the same knee joint.
(c) An accurate wear distribution map determined from the high resolution surfaces generated from the contours as in (a). (Reproduced by courtesy of P. H. Dangerfield and D. Groves.)

can be made of another new dimension, the modelling of physiological theories with computer stills and graphics. Blood flows can be photographed and pictured in this way (Plates 14, 15), and so can the operation of the gastro-intestinal and genito-urinary tracts.

Physic critically depends on the recognition of patterns of malfunction that can then be used in diagnostics (Fig. 6.14). Where the condition is so common that the practitioner can rely on encountering a case every year or two, or more frequently, diagnosis is reasonably straightforward. Where, however, the physician is dealing with an unusual case, it is immensely helpful to riffle through video-tapes as well as books, to be able to see a patient who may now be dead, or may have lived ten thousand miles away. Such capability at present depends on chance acquaintance, and who has a video-camera and is keen on using it and showing the results to visitors and students. Again, systematics in pictures cannot at present reach the level of the indexes of books.

In paediatrics, the handling of premature babies is full of atypicalities that need to be known more widely. Immune disease is also a scene of much individuality, and the spread of AIDS will require more dissemination of pictorial experience than is at present possible. Another area is the management of the condition of the growing number of the very old.

Pictures in other species; directional perception; who can see what?

Before delving into communication and recording phenomena buried in the internals of individuals, it is prudent to consider, however briefly, the possible ways in which other species may perceive pictures. We can only be speculative, because much of the solid evidence is comparatively recent. What does the brain of a bat or a dolphin do with the ultrasonic information it receives, and which we know to be a key element in its perception and behaviour? The creature reacts quickly and with sophistication. Obstacles are avoided, and there is communication with other individuals. How is the ultrasonic sensor output played into the muscle-control system, or into the consciousness? Is there some way in which a non-optical 'inner eye' puts together an image which is scanned in the way that we can see an image on the retina? We ourselves recall images in dreams and in our imagination, without the intervention of the retina: so the brain is perfectly capable of putting on a movie show with no more than its own resources.

We can ask the same questions about insects. Bees have been studied in relation to their honey-gathering and assembly. They react to colour, have direction-finding capability, and communicate through dance. Is the eye as important a carrier of their information flow as it is of ours? Some species can

Pictures in other species; directional perception; who can see what? 97

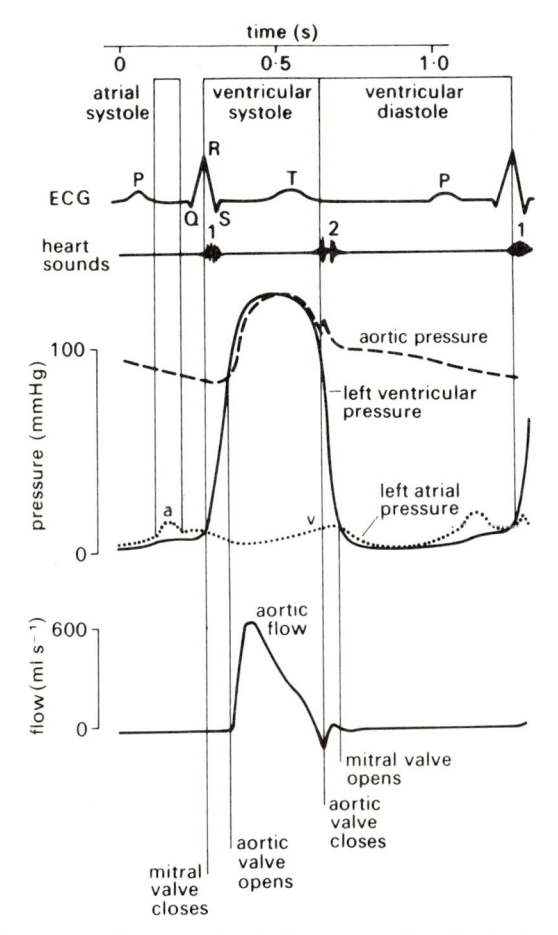

Measurement of normal function

Fig. 6.14 Electrocardiogram. Semi-diagrammatic illustration of events on the left side of the heart during the cardiac cycle. (Reproduced from *Mechanics of circulation* by courtesy of Oxford University Press.)

use electromagnetic radiation outside the optical spectrum. Is the eye the only electromagnetic detector? What maps do migrating birds have, and how do they muster the information about day-length, temperature, and other factors that determine their migratory behaviour? There has been some tendency to lump together a large number of questions under the convenient but possible diversionary word 'instinct'. How does a dog or an insect assemble its smell data? Do plants in any way survey the surrounding geography?

98   *Pictures and words in biology and medicine*

How do other species cope with the third dimension? It is commonplace to say that the siting of the eyes, with a large nose in between, interferes with the stereoscopy on which we depend. Yet many animals are no worse than we are in dealing with depth. How do they do it?

As we shall see, we ourselves have difficulty in understanding very much about our own internal recording and communication systems, to which we have direct access by our ordinary experience, and which we study intensively as part of our own medicine. It is therefore not surprising that the natural history of the internal records and communications of other species is almost a closed book, whose contents are to be guessed about by naturalists. Because of their lack of rigour, these guesses may be treated with too much contempt by professional experimental scientists. How otherwise than by guessing shall we find a way in?

The impact of biochemistry and biophysics

Our superficial glance at medicine in practice has been quite sufficient to emphasize the need for better handling and dissemination of pictorial information. At the molecular level, the roles played by the many complex giant organic molecules, in the communication and recording of information, reveal once more our dependence on pictures for understanding what happens as they react: their structures are so complex that the normal descriptions are inadequate. In this area also we encounter an unexpected feed forward as scientific perception contributes to practice of all kinds.

Even the atomic structure of the simplest organic compound, methane, cannot be fully understood if it is written with the traditional formula CH_4, nor is it sufficiently expressed as the two-dimensional structure:

$$\begin{array}{c} H \\ | \\ H-C-H \\ | \\ H \end{array}$$

The four hydrogen atoms in methane are actually arranged three-dimensionally, as the four points of a pyramid around the central carbon atom, thus (Fig. 6.15):

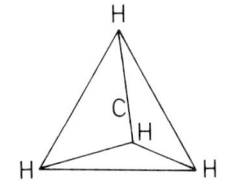

Fig. 6.15   Methane atomic structure.

The impact of biochemistry and biophysics 99

Worse, a formula that merely gives the numbers of each type of atom in the compound, as C_3H_6O, can represent more than one structure, for example in the simplest sugar, which exists in two possible arrangements. These may be described, both with alphanumeric formulae, but rearranged to express the structure, and with two-dimensional ideograms, thus:

$CH_3.CH_2.CHO$ and

$$\begin{array}{c} H\ \ H \\ |\ \ \ | \\ H-C-C-C \diagup^H \\ |\ \ \ | \ \ \ \diagdown H \\ H\ \ H \end{array}$$

(an aldehyde)

$CH_3.CO.CH_3$ and

$$\begin{array}{c} H\ \ O\ \ H \\ |\ \ \ \|\ \ \ | \\ H-C-C-C-H \\ |\ \ \ \ \ \ \ | \\ H\ \ \ \ \ \ H \end{array}$$

(an acetone)

Again, only a three-dimensional picture can reveal the full complexity of each molecule, and thus make its chemical behaviour comprehensible, for its geometry is critical to its chemical behaviour.

The twenty or so naturally occuring amino acids, which are the basic building-blocks of living cellular tissue, are linked into complex chains to form proteins. In these chains the amino acids are arranged in specific sequences, and the chains can have different forms, straight, spiral, or rolled up like a ball of wool, so giving rise to different proteins. The complexity and variety of the possible combinations of amino acids make possible the construction of very large and complex protein molecules containing many thousands of atoms. It is hardly surprising that pictures play a vital part in the process of understanding the properties of these giant molecules.

The enzyme (catalyst) lysozyme is an instructive example of such a protein (Plate 16). This is a molecule capable of promoting very specific reactions between other molecules (its 'substrate' molecules) under mild conditions. Its alphanumeric description as a sequence of amino acids gives no clue to the source of this activity, but the picture of its three-dimensional structure, as determined by X-ray diffraction, reveals a cavity capable of exactly fitting (and so activating) its substrate molecules. The whole of enzymology is now being examined in terms of this concept. Attempts are being made to synthesize simpler structures that will have similar shapes and, possibly, similar activity. The interlocking of the actions of different enzymes to produce the overall result of respiration or fermentation can be represented by diagrams of the classical chemical kind (as in Krebs' cycle, Fig. 6.16). But it is at the level of the individual enzyme that the picture is vital (Plate 17).

Similarly, the separation of reaction products by membranes and the spatial assembly of the activities of cells and organs is impossible to visualize without still and moving pictures (Figs. 6.20a and b, p. 108). Electron and optical

Classical chemical diagram

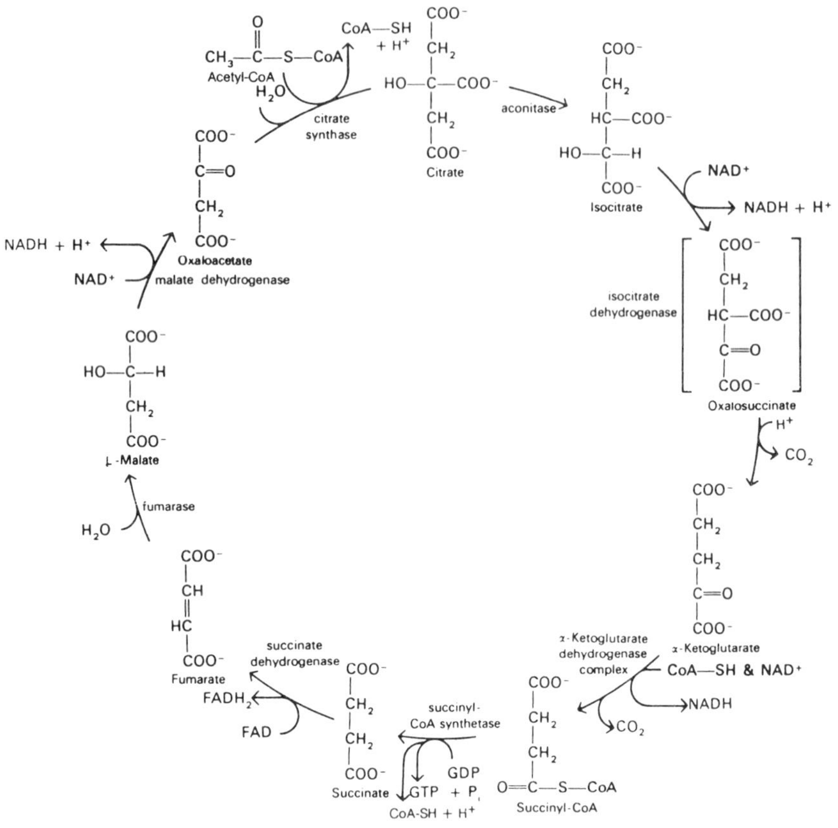

Fig. 6.16 The Krebs cycle shows the complexity of chemical reactions in respiration. (Reproduced from *Biochemistry* by courtesy of Oxford University Press.)

microscopy reveal enough about the topography of nerve networks (Fig. 6.20, p. 108) that, today, with knowledge of the unit processes of nerve action, we may begin our understanding of nervous systems of many kinds, and start speculation about brain and optical mechanisms.

Key elements in the work of Crick and Watson on the DNA molecule, in the 1950s, were their models and inspired guesses about types of structure likely to be of significance in the actions of nucleotides, and nucleic acids, as these guide protein synthesis. A conviction that there was fundamental importance in the spirals of Linus Pauling's simple peptide structures led to the interpretation of existing X-ray diffraction patterns of bigger biologically interesting molecules in terms of spiral models. The central part of the research consisted of the building of plausible, real, three-dimensional atomic models that stood on the laboratory bench (Plate 9). Only when a satisfactory model had been found, which explained some curious chemical regularities, was it seen that the emergent structure was a piece of machinery for exact self-reproduction. There were two spirals linked by chemical bonds (Fig. 6.17), which could be formed and broken in patterns that allowed each of the spirals (or helices) to act as the template for new spiral-building. Once more, pictures and models provided the clues. The complete cycle of reactions in cell reproduction embodies more imagery then alphanumeric abstraction.

Yet the DNA structure has a significance for us that goes beyond the need for sophisticated imagery. It illustrates in a vivid fashion the enormous ability that large organic molecules have for storing, sorting, and transferring information. For this reason we need to take a closer look at this giant structure.

Genetics and its cryptology and xerography

When a cell divides, each chromosome within it must also divide so that each daughter-cell will carry a full set of chromosomes identical with those in the original cell. This replication is necessary for all growth, whether from fertilized ovum to fetus to childhood and subsequent adulthood, or from acorn to oak tree, or for repair and renewal of tissue. The structure of the DNA molecule in each chromosome makes this replication possible. The molecule consists of two strands, made up of smaller molecules, intertwined in the form of a double helix (Fig. 6.17b). Each strand is constructed of a sequence of alternating sugar and phosphate molecules. Attached to each sugar is one of four nucleic acid bases, thymine, adenine, guanine, and cytosine. Each base is paired with another base in the adjacent helical strand (Fig. 6.17a). This linkage is so organized that the base thymine is always linked (via a hydrogen bond) to an adenine in the adjacent strand, and cytosine always to guanine. Thus the sequence of bases on one strand determines precisely the sequence on the other. This means that when the two strands separate at the beginning of cell division (breaking the hydrogen bonds), each separated strand acts as a template, carrying coded information for the chemical synthesis of a new helical strand with which it will intertwine (Fig. 6.17c). The question of accurate replication is, however, only part of the story of information-processing in DNA. It also has the ability to pass on instructions to other molecules.

Crick, Watson, Perutz, Dorothy Hodgkin, and others provided the cave-paintings of molecular biology, which we have just looked at. What is truly

102   *Pictures and words in biology and medicine*

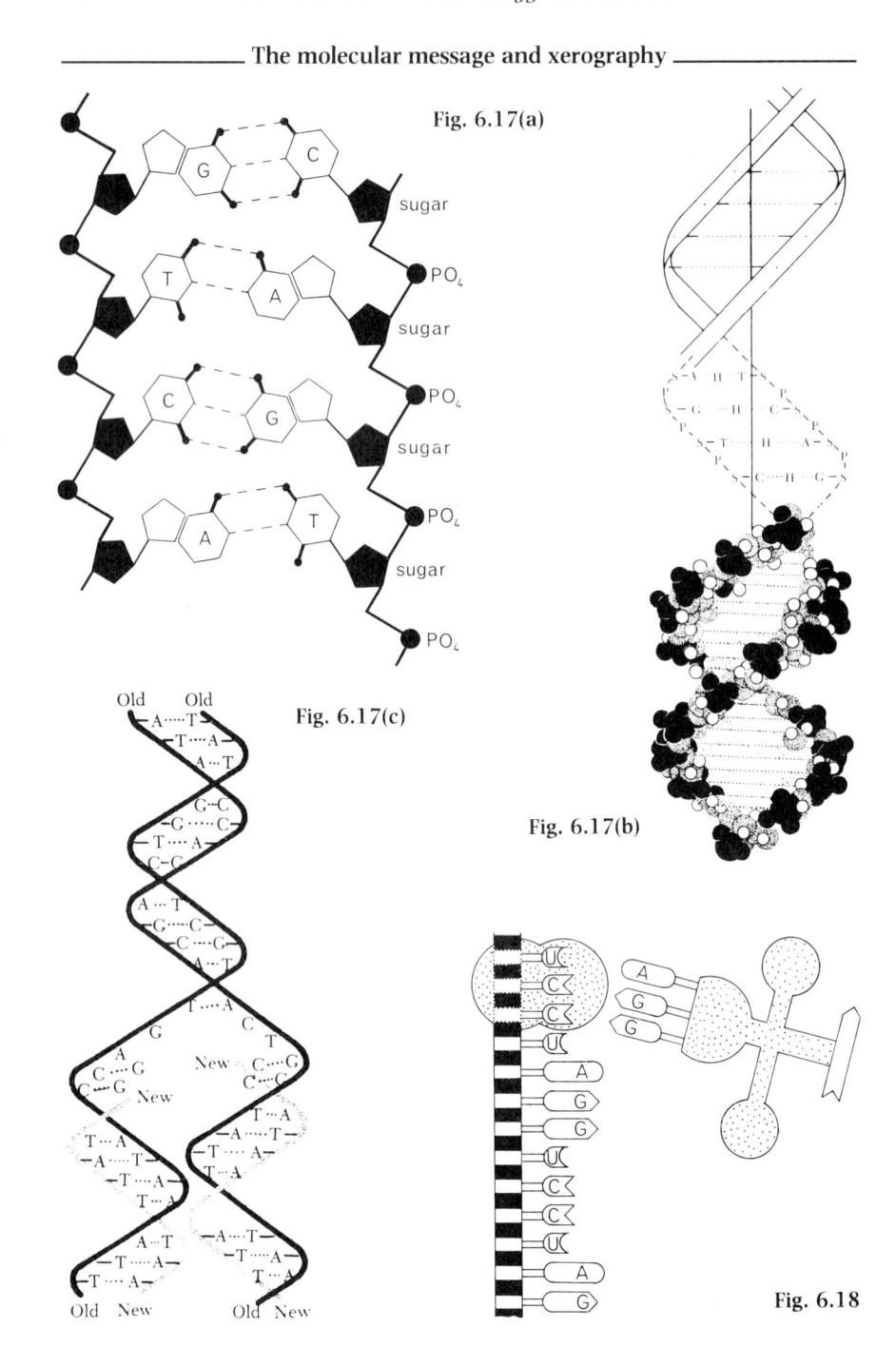

────────── The molecular message and xerography ──────────

Fig. 6.17(a)

Fig. 6.17(c)

Fig. 6.17(b)

Fig. 6.18

astonishing is that, in the case of DNA, we did not have to wait for 300 centuries for the alphabet and syntax that enabled the new knowledge to be codified and used. During the 1960s it became clear, as Sir Edward Pochin (1983) neatly puts it, that 'the sequence of the four bases in these molecules constitutes a code which tells the cell what to do, and what chemical substances to produce. The molecule is the message'. The breaking of this genetic code was achieved when it was realized that the bases are arranged on each strand, and must be read, in groups of three, each triplet providing a single word of the genetic code. By this device, the four letters GCAT (derived from the four bases) can give rise to 64 different words (four possible variants in each triplet, thus $4 \times 4 \times 4 = 64$).

The significance of this discovery has, again, been nicely expressed by Pochin in the same work. He wrote that this number 64 'is more than enough, because the function of the vocabulary is, essentially, to spell out the way to construct all the different proteins which the body contains and uses. Each protein consists of a long chain of amino-acids, and the body ordinarily uses only 20 kinds of amino-acids in its proteins. A vocabulary of 64 words is therefore fully adequate to code for each amino-acid and indicate the sequence in which they must be assembled in synthesizing the different proteins for which the various chromosome pairs are responsible. Indeed, several amino-acids are coded for by as many as 6 different triplets. This lavish allotment of different triplets to code for the various amino-acids, which undoubtedly is saying something about our far evolutionary past, still allows spare triplets as punctuating instructions, for example, to say "Start reading from here" or "Stop. Protein completed", and so to indicate where to start reading each triplet. This is crucial, as any cryptographer knows, since a cryptogram written in blocks of three can make a very high grade of nonsense if begun at the wrong point (an dan yon eca nse ewh y)' (Table 6.1).

Fig. 6.17 The giant DNA molecule.
(a) The arrangement of phosphates, sugars and bases, A with T and G with C, held together respectively with two and three hydrogen bonds. (After P. C. Hanawalt and R. H. Haynes, *The chemical basis of life*.)
(b) The DNA helix, shown in three ways. At the top, the general arrangement; in the centre, general detail of chemistry; at the bottom, how the atoms are placed. (Reproduced from *The cell* by permission of Prentice-Hall Inc. Englewood Cliffs, New Jersey.)
(c) The helix unwinds to form two new helices. (Reproduced from *The cell* by permission of Prentice Hall Inc. Englewood Cliffs, New Jersey.)

Fig. 6.18 RNA messenger vehicle engaging with the appropriate triplet. Note that in the RNA, uracil(U) takes the place of the DNA thyamine(T).

Table 6.1
The genetic code

	U	C	A	G	
U	phe	ser	tyr	cys	U
	phe	ser	tyr	cys	C
	leu	ser	*punc.*	*punc.*	A
	leu	ser	*punc.*	trp	G
C	leu	pro	his	arg	U
	leu	pro	his	arg	C
	leu	pro	gln	arg	A
	leu	pro	gln	arg	G
A	ile	thr	asn	ser	U
	ile	thr	asn	ser	C
	ile	thr	lys	arg	A
	met	thr	lys	arg	G
G	val	ala	asp	gly	U
	val	ala	asp	gly	C
	val	ala	glu	gly	A
	val	ala	glu	gly	G

The Genetic Code The twenty amino acids (names abbreviated) which make up the body's proteins, fitted into a table of the genetic code. Of the 64 possible encodings note that three are for punctuating instructions. (After D. R. Hofstadter *Gödel, Escher, Bach: an eternal golden braid*.)

He goes on, 'We therefore have a basis for reading off the recipe required for constructing a complex protein out of the right sequence of the available amino-acids. The cell has its own procedure, of reading the recipe in the library of its nucleus, and making the protein in the kitchens of its cytoplasm. It copies the sequence of bases recorded on the chromosomal DNA (deoxyribonucleic acid) by forming molecules of the chemically rather similar RNA (ribonucleic acid), using the DNA as a template so that the sequence of bases on the RNA forms a counterpart of that on the DNA.' [There is a change in the coding in RNA because, as seen in Fig. 6.18, the thymine of DNA is replaced in the RNA by uracil, so that T is transcribed as U, but the code remains.]

'The strands of this "messenger" RNA then migrate out of the cell nucleus, and carry the copy of the recipe into the ribosomes in the cytoplasm of the cell. The various ingredient amino-acids arrive, each kind being carried on its own specific vehicle in a manner reminiscent of the traditional Hindu pantheon, except that here the "vehicles" are short sections of RNA. These molecules of "soluble" RNA each engage with the appropriate triplet on the recipe in the ribosome, so that the various amino-acids are literally lined up in the correct sequence, to correspond with the sequence of bases in the original chromosomal code.' (Fig. 6.18). Organic molecules have, therefore, the capability to sort information, to replicate it, and to transfer it.

A fourth factor of importance is the immense quantity of information that is dealt with. The total length of coiled DNA in a human cell is about 2 metres. It contains about 8 billion paired nucleic acid bases, shared among the 46 chromosomes.

If the rapid solution of the genetic code was astonishing, no less so is the fact that so much of the primary process of life depends on linear molecules (such as are found also in the thermoplastic polyolefins) whose description needs only linear words. The basic chemistry could so easily have involved cross-linking (like that in the thermoset resins) whose description needs acrostics rather than words. A typical protein is made up of a chain of some hundreds of amino acids arranged in particular sequences, as shown earlier in the structure of lysozyme (Plate 16). It is now understandable that such a structure can be encoded by using the triplet codes of the amino-acids. This means that this information is easily indexed and sorted by our ordinary and familiar lexicographic procedures. Molecular biology has walked straight into the welcoming arms of the pre-existing digital computer, which is ideally constructed to process the large banks of information now accumulated on DNA and protein sequences. Biology, having been (and still being) so dependent on pictures (Plate 18), turned out to be a great place for using a quite new alphabet and language comprising very simple words.

It may be that we should not lose our heads and assert that our luck is all-embracing. There *are* cross-links in proteins, so that in the word-patterns from which we try to predict the *shape* of enzymes, there are subtle and crucial acrostics. Moreover, some of the words describe operations rather than amino acids (such as the 'start reading from here', etc.), and also there are minor variants of the pyramidines (thymine and cystosine) and purines (adenine and guanine) that may require more letters. Thus pictures may be a feature of symbolic biology, as represented by the more refined genetic codes to come, as well as an essential element in the description of biology as it is seen by the eye or the microscope, or in the tomograph.

Pictures and words in biology and medicine

Novel, pictorial support for the Darwinian theory of natural selection

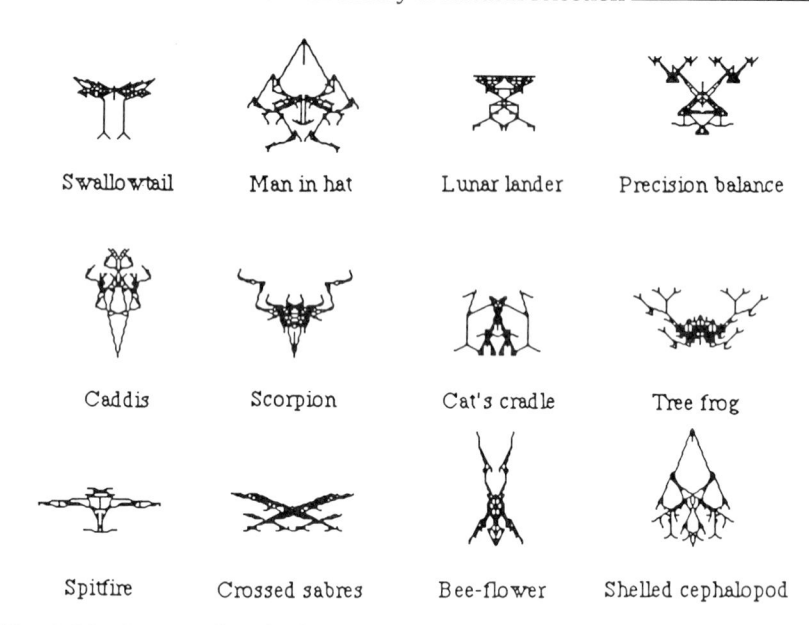

Swallowtail   Man in hat   Lunar lander   Precision balance

Caddis   Scorpion   Cat's cradle   Tree frog

Spitfire   Crossed sabres   Bee-flower   Shelled cephalopod

Fig. 6.19 Biomorphs which evolved inside Dawkins's computer when it was programmed to generate forms accumulating small changes. (Reproduced from *The blind watchmaker* © Richard Dawkins 1986 by courtesy of Longman Group UK Ltd. Permission from Peters Fraser & Dunlop.)

Such need for caution as to detail is reinforced by elementary thought about reproduction. Integrity of protein and DNA identity is an essential foundation for replication, but we know so little about developmental biology and the control mechanisms governing the construction of cells, and their assembly into organs and other components, that we ought not to assert that linear words are the only carriers of information. The double helix plays the part of the central item in biological xerograhy (replication). The copying process is notably precise, and may be more complex than is apparent. Such precision is essential. The quality control necessary for offspring to be viable and healthy is at a level many orders of magnitude greater than anything known in the artificial world. Dawkins, in *The blind watchmaker* (1986), has likened it to the correct copying of a document, where one typist copies from another who copies from another in a sequence of 20 billion successive accurate copyings. Further, the information flux in such xerography would seem to be well over ten to the power of 30 bits/sq cm/sec, which again is far beyond the best of our technologies. The fertilized ovum that is going to produce an Albert Einstein has to divide with a fidelity that makes possible transfer of all the necessary information (at least ten megabytes in computer terms) across a small boundary within hours. The linear speed of transfer of information is small, but the density is immense. Morever, the improvement of species and the generation of new species occur by accumulation either of beneficial errors or of adaptive modifications. The language of such incremental change may be simple and Darwinian, as expressed powerfully and graphically by Dawkins's biomorphs (Fig. 6.19), or it may be more complex. Accordingly, we should be prepared for pictures of even greater complexities, to express and manipulate the relevant knowledge when we have it.

Brain science; artificial intelligence; molecular memories

We now enter the most difficult topic in the entire book. What is the language, and the intellectual framework, which we can most usefully employ in understanding the brain and in building machines or systems that will mimic some of its simpler capabilities. As observed earlier, the amount of information stored by the brain, the speed of its transfer, and the sophistication of its sorting and switching, is already exceeded by digital computers. Yet the brain's cryptology gives it a key advantage. It is better able to handle and compare information, which is stored in a form much closer to that in which it is received, than are the present machines. It may be that everything that the brain can do is explicable in terms of the recognition and manipulation of complex patterns and packages of information. A composer does not assemble notes. He assembles phrases, cadences, sounds, and emotions. He plagiarizes from himself as well as from others; but in doing so he enables his hearer to make cross-references in his own memory. Sometimes these references are explicit, as in the quotations from other composers in Richard Strauss's *Capriccio*. Sometimes they are contextual, as in the *Leitmotiven* of Richard Wagner. A two-chord sequence makes the listener think of Fate, a five-note phrase recalls the Sword. These elements—a hundred or so of them—weave in and out of the music without interrupting it. Sometimes they are obvious, and sometimes deeply concealed. The brain, like a dog rooting for a smell, can find and retrieve them without stopping the process of enjoying the music or appreciating the quality of the singer, the setting, the production, or any of the instruments in the orchestra, not to mention the influence of the conductor.

Neurologists and others have mapped the brain (Plate 14, Fig. 6.20d). They have recorded many occasions when, following a stroke or other calamity, functions have been transferred adaptively from one zone to another. Psychologists have studied memory and the ways in which it can be trained or

108   *Pictures and words in biology and medicine*

Nerve networks and their transmission system

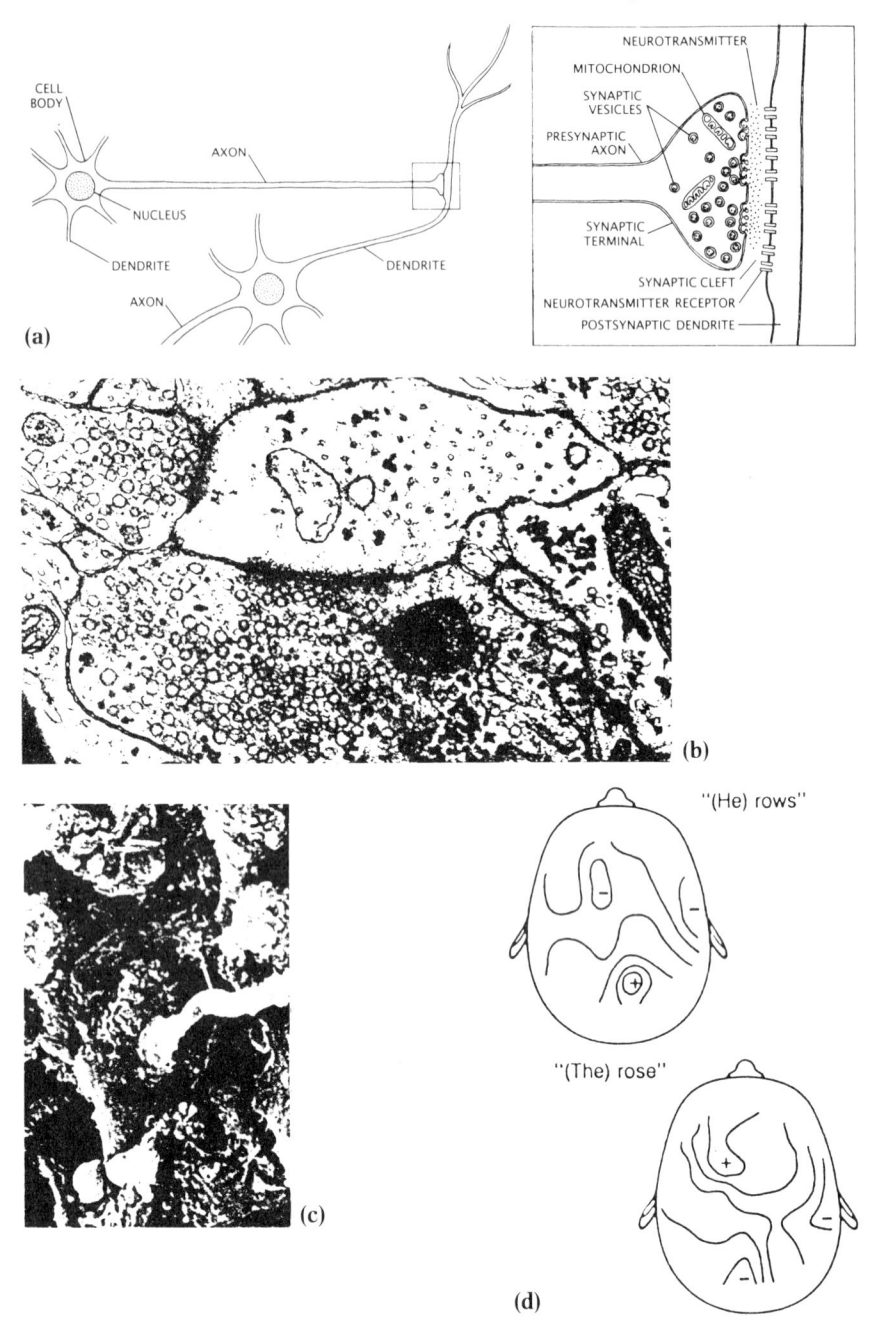

(a)

(b)

(c)

"(He) rows"

"(The) rose"

(d)

Brain science; artificial intelligence; molecular memories

impaired. Others have explained some brain processes in terms of networks, following experiments on species such as the octopus or the dolphin. Most investigators believe that the explanation of brain function involves the understanding of parallel processes in complex networks (Fig. 6.20). They feel that parallel processing computers may be better than vector processors—however complex and advanced these may be—in mimicking thought. Their use, however, requires the writing of a considerable amount of software of a new kind.

Meanwhile, the software writers are producing 'expert systems' which try to imitate some of the end-results of mental activity. If we can set up a machine which conducts intelligent and friendly medical diagnosis on the basis of questions to the patient, coupled with the surveying of large tracts of medical information in the critical way that a skilled physician would survey it, and if this can usefully supplement (and sometimes replace) the work of a physician, there will be two results. First, we will have a useful new ability. Secondly, we may learn something about the way in which a human expert uses information where there are some internal conflicts. Even if we do not find out immediately *how* the brain does the job, we may learn more about *what* is being

---

Fig. 6.20 Neurons, the brain's basic hardware, communicate at specialized junctions called synapses, inducing graded electrical potentials in the receiving cells. These potentials, summed over thousands or millions of neurons, give rise to a slow electromagnetic wave, or EEG, that pervades the brain.
(a) Diagram showing how the neurons transmit signals at the synapses. A minute electrical disturbance (the action potential) travels from the cell body down the axon. It reaches a bulbous structure, the presynaptic terminal. Vesicles in the terminal release a neurotransmitter (amino acid) that diffuses across the cleft and binds on to the receptive fibre (dendrite), eliciting a new electrical signal. (Reproduced from GABAergic neurons by courtesy of *Scientific American*, **February 1988**.)
(b) High-power view of two nerve cells contacting a dendritic spine. The synapse shows clearly as the darker, thicker region of membrane which separates them, and the circular vesicles are seen clearly. (Reproduced from *Essays on the nervous system* by courtesy of Oxford University Press.)
(c) Scanning electron micrograph of neurons. Synaptic terminal and axon visible. (Reproduced from *Principles of neuroanatomy* by courtesy of Oxford University Press.)
(d) The brain-wave patterns produced by listening to the verb 'rows' and the noun 'rose' differ. The diagram shows the averaged EEG patterns based on records from many people. (Reproduced from Why can't a computer be more like a brain? by courtesy of *High Technology*, **August 1984**.)

done and achieved. In due course, such pragmatism may give leads on the 'how' question.

Such practical artificial-intelligence packages as expert systems can cover many tasks: oil exploration, weather forecasting, obscure fault-finding, market analysis. The most successful examples to date are those where the information processed is entirely linear, that is, verbal or numerical. However, even simple and not very expert tasks, such as the question whether two photographs are or are not of the same person, or whether any of the participants in an identification parade were the same person seen in the street on a particular day, are difficult to handle by computer. The machine can provide processes for synthesizing images and modifying them until a human at the controls believes that he has produced a good and accurate image of something or someone seen. (In the Appendix Professor Meadows explains the application of parallel computers to this problem.) As with sorting and indexing, the machine is in more trouble with the analysis, although some progress has been made in the automatic detection of linear features, such as asbestos fibres, cracks, and blood vessels (Fig. 6.11, p. 93; Plate 19).

Examples may help. One of the best areas for the computer is that of games like chess, with precise rules. All the possible courses for some end-games, with very few pieces on the board, can be tabulated in detail by the computer. It can run through all the consequences of all of the possible moves, starting from any position, and pick the one with the best prospects, repeating the analysis after the opponent's next move. This unintelligent process proves to be highly effective, since an experienced chess player cannot survey so exactly all the possibilities, and falls back on 'heuristics'—rules of thumb that reduce the number of cases to be thought through. In the case of the King and Queen vs King and Rook end game, the player with King and Rook will normally keep his two pieces together for mutual protection against the superior attack. This denies him some strategies which the computer, with its look-up table, is able to select with confidence, if they are assessed as better. So the computer, playing with the inferior position, is usually able to prevent a highly-skilled Master from winning with King and Queen. A human Master playing King Rook will more often get beaten. Again, the simple computer strategy 'play to optimize the results of exchanges of pieces over the next three moves, and thus avoid silly mistakes' does better than many very skilled attack strategies. This is one of the most disappointing of computer findings. Chess is now seen as a game that penalizes mistakes more than it rewards brilliance.

Pictorially based judgement is most prominent in the analysis of satellite telemetry to find particular objects. The search may be for particular shapes and outlines (airstrips, missile launchers) or for particular kinds of area, that might be ore-, oil-, or water-bearing, or for particular types of cloud. As noted in Chapter 4, research on pattern recognition is partly in the science sector and partly military. With present computers, the methods are clumsy.

This raises the question of new computers. Parallel processing is one option (Appendix A). It is a rather slow one to get going, for the reasons given earlier. A more radical approach is to try to invent new types of memory, including chemical types of the kind which might emerge from biological evolution (say, in large molecules). Knowledge and experiment on the handling, sorting, and analysis of pictures may well be a useful component of research into the working of the brain, and *vice versa*. The biotechnology community has now identified just this subject as one of its key activities in the years to come, and it is important to accept that the need is not just to find ways of applying the most advanced conventional machines to defined biological problems (for example, databank sorting, or control engineering), but to design new hardware and software on the basis of biological perception. Once more, pictures will be important as records and carriers.

7. Pictures and words in everyday life and study

Background: the pervasiveness of television

We have reviewed the slow fall and the recent rapid rise in the importance of pictures throughout the last three hundred centuries, in education, science, engineering, and medicine. Now we need to turn once more to the man and woman in the street and the supermarket and enquire where all of this has left them. He or she may, of course, be the modern counterpart of Molière's *Bourgeois Gentilhomme*, who was astonished to be told that he had been speaking in prose for the whole of his life. Our ordinary citizens have only recently become habitual and sophisticated picturegoers and film critics, and may have only a dim recognition of the depth of their involvement with pictures.

A survey of the total effect of television is beyond the scope of this short book, but some points are worth making. The immediate pictorial presentation of human catastrophes from all over the world has an impact different from that of the solemn printed word in the newspaper, with the occasional photograph. If 'any man's death diminishes me, because I am involved in mankind', then the sight of so many deaths, while they are actually happening, might indeed be daunting. We could be permanently borne down by the burden of the bell tolling for us, since death or discomfort are the centres of so many news stories. The preservation of sanity demands, accordingly, the development of insensitivity. The spectacles of wealth or poverty on an unfamiliar scale, which might provoke crippling envy, complacency, or concern, appear to give rise, in fact, only to enhanced but muted aspiration and concern. In a different area, the display of accessible affluence helps the advertisers. Entertainment may be degrading or enlightening: most of us get some of both. On the positive side, television again and again opens doors: we suddenly see worlds of which we had been unaware, and in which our continuing interest is aroused. The universe, the galaxy, stars, planets, rocks, molecules, atoms, paintings, sculpture—all swim into our ken. Visions of the microscopic appearance of familiar solids or organisms, mechanisms of engineering, the workings of the brain, the concepts of science, and the subject of *The telling image*, all come crowding in through their televisual introduction.

In this chapter we can do no more than look briefly at some particular pictures—stylized, formalized, or otherwise—which can be specially identified as important in principle. First, however, we must follow McLuhan, and look at the medium itself, which is the message.

How to display a picture

We may look at a picture in two, three, or four dimensions—flat, solid, or moving (time being the fourth dimension). We may also add other dimensions of an imaginative kind. In his composition *Pictures at an exhibition*, Mussorgsky used his imagination to give pictures a musical dimension. Wagner's *Ring* has even more dimensions of this kind—depth, time, music, speech, and intense perception and prejudice—eight dimensions in all, including the two of the flat picture. The medieval passion play, performed on a cart in the market place, had minimal support. Whether it displayed more or fewer dimensions than the tragedies of Sophocles or Aeschylus raises an interesting discussion. In the Prologue to *King Henry the Fifth* Shakespeare invokes the imagination explicitly: 'Think when we talk of horses, that you see them. Printing their proud hoofs i' th' receiving earth'. Words themselves are multidimensional. How many dimensions are there in the lines starting, 'I met a traveller from an antique land' or 'Much have I travell'd in the realms of gold'?

High technology is not essential for imaginative extra dimensions; but it helps. An audio-tape machine can be hired to hang round your neck and guide you through the art gallery with the spoken word. On a television screen we may engage in montage, and insert or superimpose other pictures or words. We can place the verbal text on screen and see the coloured picture through and behind the letters.

One of the more immediately and obviously useful aids to understanding provided by technology is the caption in words. Pictures have always had descriptive titles such as *The Scapegoat* or *Rain, steam, and speed*, but the television view-data subtitles for the deaf, or snatches of opera libretto in the domestic language, can go further, and penetrate the bewilderment created by lack of hearing or of understanding of a foreign language. Another valuable aid is software, such as that supplied by Quantel, which can show an inset picture of the manager watching the football game, or a comparison between two pictures being faded out and in. The variable-speed playback enables the armchair observer to see the game better than the poor beleaguered referee. Whether all the tumbling, drawing, and quartering always helps, is a question which might be argued.

114 *Pictures and words in everyday life and study*

───────────── **Pictures and telephones** ─────────────

Fig. 7.1

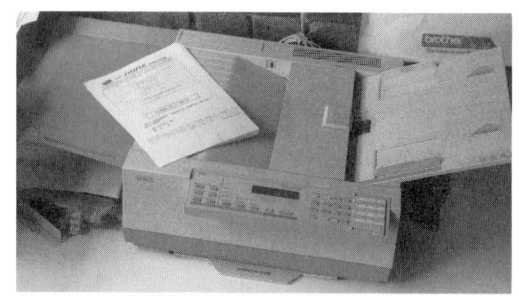

Fig. 7.2

Fig. 7.3

The specific benefit of depth (the third spatial dimension) has already been spoken of in Chapter 6 in connection with enzymes. The fourth dimension, of time, is implicit in the structural regularities of the DNA double helix, which is specifically adapted for change. So far, three-dimensional television, based on light polarization or holography, is not generally available.

Perhaps the most interesting case of all is the use of the picture as an aid to the telephone, which, at present, remains restricted to sound, the first video-phones having been a flop. However, the simultaneous use of the copying facsimile machine (Fig. 7.1) and voice only requires one more pair of phone lines to give important extra perception at little additional cost. By contrast, video-conferencing (Fig. 7.2) needs a bandwidth equivalent to 50–100 telephone lines. Nevertheless, regular video-conferencers speak highly of the method and of the saving of the time, expense, and the fatigue of travel. It may, however, puncture the self-esteem of those business travellers who dash about doing little more than being busy and providing themselves with an excuse for not thinking. Other conferencing aids are the electronic writing board and pad (Fig. 7.3), which are used as visual aids via a telephone link. Anything written or drawn on the board is reproduced not only on a monitor in the same room, but also on a similar display elsewhere. Conference members may augment or change the display by linking up their own board or pad. There is also electronic mail, the use of which marks a trend towards the paperless office. Video-interaction will become cheaper, and available anywhere: in the Oval Office, the amphitheatre at Verona, or on top of Everest. In due course, we shall be able to enjoy and be members of conferences of holographic ghosts, who will be able to shoot, kick, or punch one another, or crown one of their number with the Triple Crown.

The return of pictograms and ideograms

Alongside the massive growth in the use of representational pictures to convey and record information, and of precise schematic computer-drawings in

---

Fig. 7.1 A facsimile machine capable of reproducing pictorial images.

Fig. 7.2 Video-conferencing. Note too, the graphic display of information. (Telefocus; a British Telecom photograph.)

Fig. 7.3 Electronic writing board, big brother to the writing pad. Four colours are available, and distant attenders see as well as hear the presentation. (Telefocus; a British Telecom photograph.)

engineering and elsewhere, there has also been a considerable expansion in the use of simple pictograms and ideograms for labelling and instruction (Fig. 7.4). This is partly because international communication, travel, and sale of goods have been growing rapidly. Traffic signs, the coded washing instructions fixed to garments, and pictorial instructions for the assembling of equipment are typical examples. Such signs cross language barrriers, so that the driver of a car receives the same emergency or directional indications in countries with different languages. They tell him where he may not park, and how fast he may move. They play a prominent role in road safety. In providing guidance, such as where to turn off for Toller Fratrum, they not only help the driver to find his way, but give him confidence, and thereby reduce tension. The roundabout sign in Fig. 7.5 is, in fact, a lie, since the circle of road is not broken; yet the message is clear that, to get to Yr Wyddgrug, one may not turn right, but must continue clockwise. All of these symbols allow a degree of rapid comprehension of different bits of information simultaneously in a way that would be impossible if the notices were designed with letters and numbers that had to be read in sequence.

At the same time there is a continuing search for words that convey roughly the same message in different languages. These are then suitable as international trademarks, indicator signs, and product names. The fuel gauge of a car made by an international company may be indicated by a little pictogram of a petrol pump, or by the international work TANK, which is preferred to the words petrol, *essence*, or *benzin*, each of which has currency only locally. Car models are named Golf, Polo, Metro, Maestro, Accord for the same reason. Computer and instrument instructions are also being tried out with pictograms instead of words (Fig. 7.6). This avoids the need for different routines for different countries. Whether such graphics simplify the process is doubtful. But the use of graphic displays to show the state of a system can be helpful in making it easier to concentrate on priority messages, and to avoid being confused by multiple danger warnings issued as a variety of dial readings, red lights, or sounds. We shall deal with this matter more fully in a later section.

Alongside the explicit pictograms there are ideograms which act as logos for firms and public authorities. These are believed to help sales or comprehension once an organization has established a good reputation. They can be used as a 'seal of good housekeeping' to help launch new products, services, or other initiatives (Fig. 7.4). An important and long-standing use of ideograms is for military recognition, so that members of the same unit can recognize their fellows with whom they are meant to be co-operating, and distinguish them from others. These symbols are then used to encourage *esprit de corps* through their display on flags, banners, trucks, aeroplanes, and so forth. Companies sometimes employ their logos in this way: a good 'house style' can have much merit.

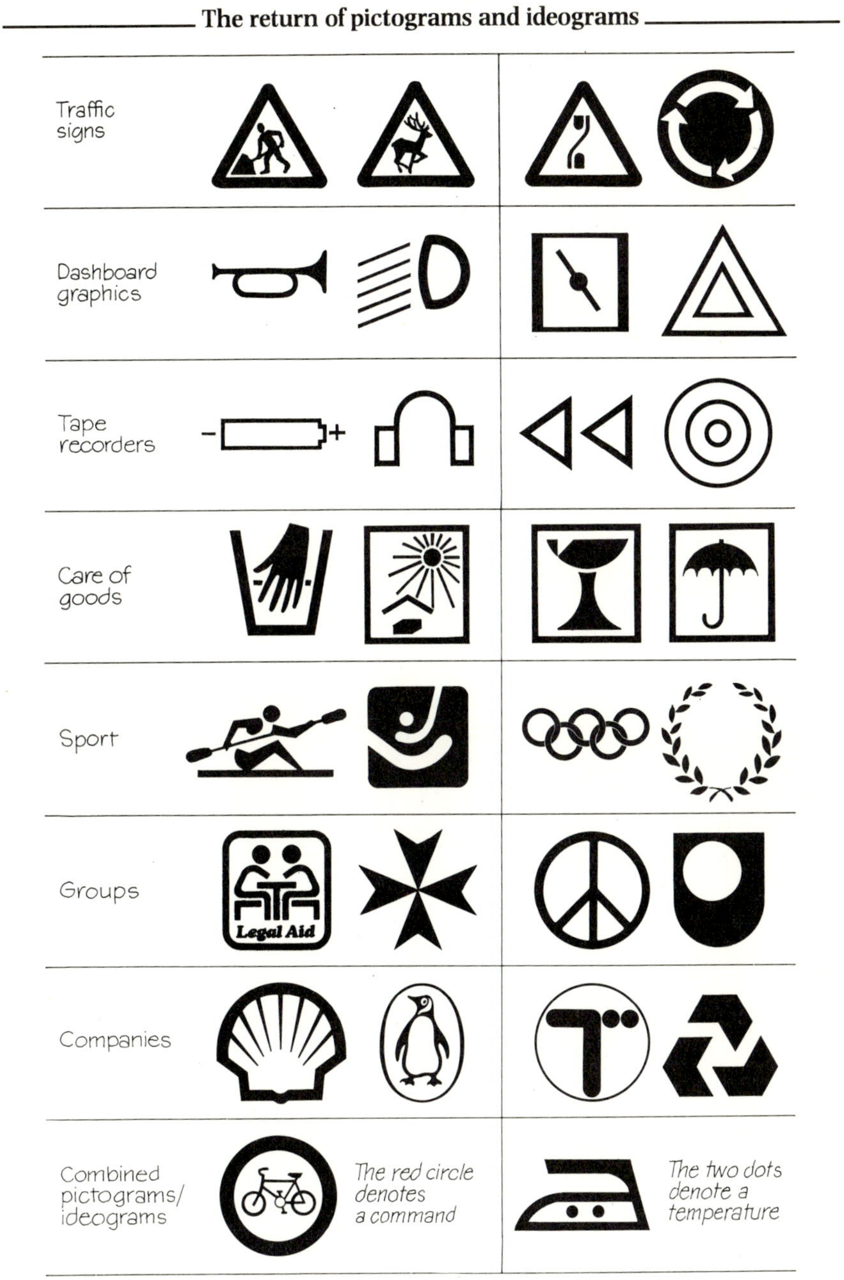
The return of pictograms and ideograms
Traffic signs
Dashboard graphics
Tape recorders
Care of goods
Sport
Groups
Legal Aid
Companies
Combined pictograms/ ideograms
The red circle denotes a command
The two dots denote a temperature

Fig. 7.4 Modern pictograms and ideograms. (Reproduced by courtesy of Fraser Graphics.)

118   *Pictures and words in everyday life and study*

————————— **The return of pictograms and ideograms** —————————

Fig. 7.5 The broken circle of the roundabout. Is this a kind of convention which will be used elsewhere?

Fig. 7.6 Pictorial computer symbols.
(a) Symbols on the Daily Appointments page.
(b) Apple HyperCard tools palette with added key.
(Reproduced by courtesy of Apple Computer UK Ltd.)

(a)

(b) Browse   Button   Field

Rubber ——————————————— Lasso
Selection ————————————— Pencil
Brush ————————————————— Line
Spray —————————————————— Rounded Rectangle
Bucket ————————————————— Curve
Rectangle ————————————— Oval
         Paint Text ———   ——— Polygon
              Regular Polygon

Maps

Our prototype citizen has an introduction to geometrical projection through his use of maps. If he is already a hill walker, an offshore sailor, an enthusiastic motorist, a surveyor, a geographer or a geologist, maps will have been part of his life for a long time. The oldest use of maps was probably by armies: battle plans, the choice of ground, and the writing of histories all depend on maps. Armies more often than not are responsible for the preparation and maintenance of the maps that everyone uses. Hence the name 'Ordnance Survey' in Britain. Navies are even more explicitly involved in charting and exploration: the third dimension, and the location of deeps, shallows, and channels matters seriously to them. The use of the digital computer, with its flexibility, has speeded up the preparation and amending of maps, and made easy the integration and conversion of data of many different kinds (Plate 20).

The man in the street may have learnt about maps as a member of a wartime army, a volunteer reserve, a conscripted peacetime army, or the Boy Scouts. So it is perhaps unlikely that he will ever have been in the position of the Liverpool girl whose friend, on hearing that she had been on holiday in the Seychelles, asked 'Where's that?', and received the answer 'I dunno: we flew'. Nor (to move up market but not up intellect) will he have played the part of the businessman in the totally standard hotel, who had to ring up his secretary at base to ask 'Where am I?', to be told 'It's Tuesday, so you are in Brussels'. He will have seen quite a number of maps on television, some of which are now in the field of entertainment. Not long ago, verbal weather forecasts were heard, often accidentally, and then complained about. No one had much appreciation of the reason for the talk about 'a deep depression over Iceland' ('Where is Iceland?'). Now, however, maps and satellite-picture animation and simulation, together with some imaginative casting, have changed all that. People understand something of the craft of weather forecasting, and watch the forecasts voluntarily and even eagerly. Many more people know the location not only of Iceland, but also of the Faroes, and even of America and continental Europe, which also appear in the supporting evidence (Fig. 7.7). News bulletins include maps showing the location of Soweto, test sites in Nevada and Siberia, the routes of airliners which are buzzed, hijacked, or worse over Korea or Syria, or the best way up Everest. We have seen maps of the Moon and of Jupiter, and even of the great galactic spiral. Consequently, we have not only a substantial amount of understood new knowledge, but an appreciation of the utility of projections on a plane (Fig. 7.7).

Minds are therefore prepared for an engineering or an architectural drawing. All of this education has been achieved without public expense or the personal tedium of preparation for GCSE Geography. The only price has been a few minutes less of watching inane chat or quiz shows. It would be interesting

120   *Pictures and words in everyday life and study*

────────── **Geometrical projection and the weather map.** ──────────

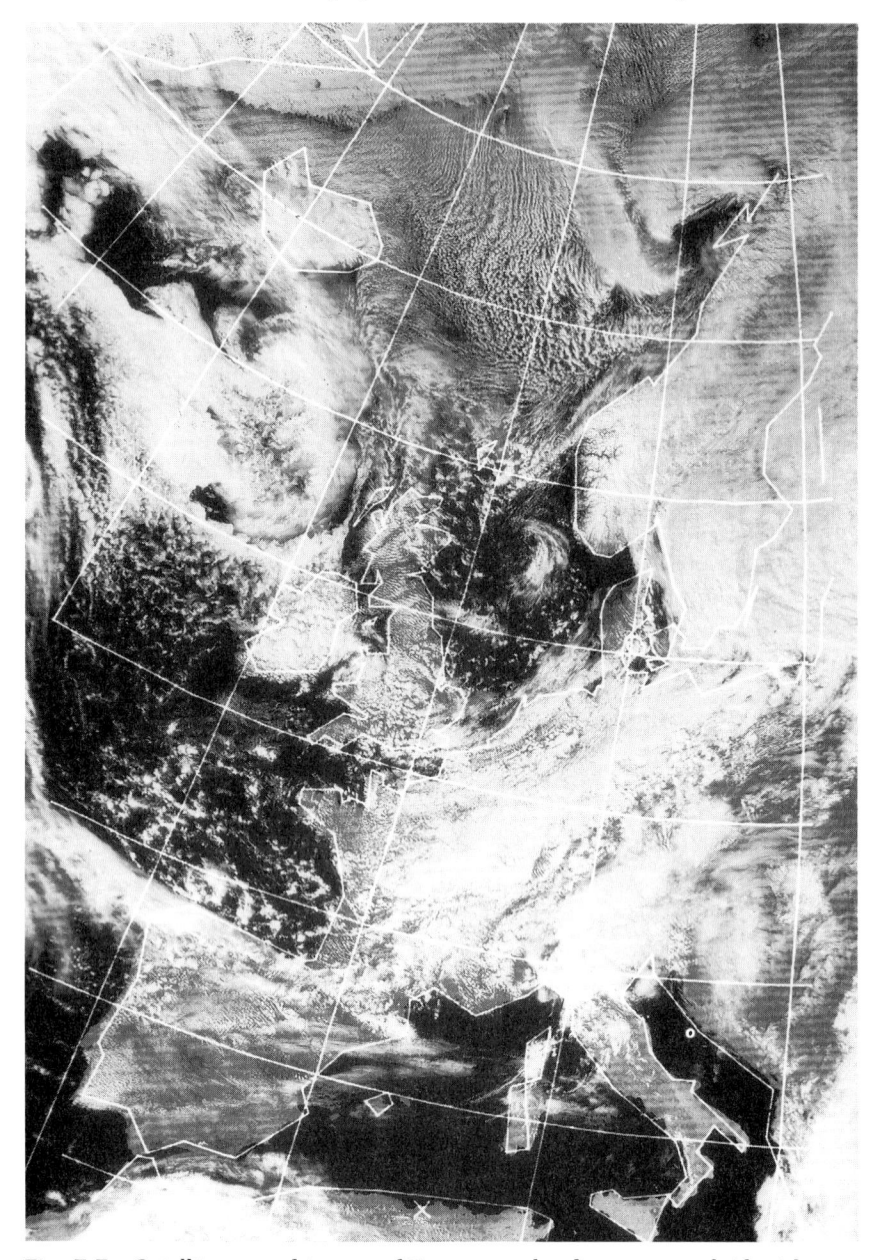

Fig. 7.7 Satellite-sensed image of European cloud cover, overlaid with map projection. (Reproduced by courtesy of the University of Dundee.)

Maps 121

_____ **Mapping by satellite** _____

(a)

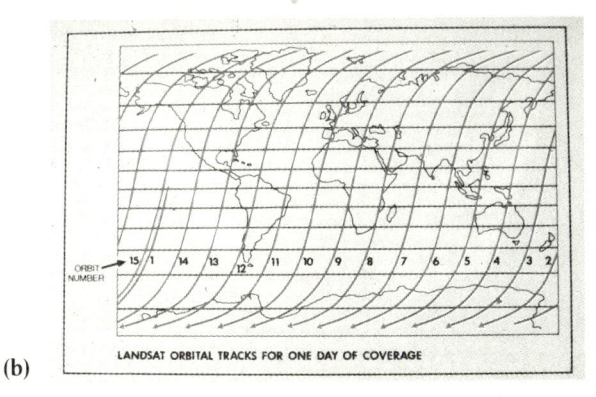

(b)

Fig. 7.8 (a) The satellite *Landsat* and (b) its route map: one of many satellites from which information is continually telemetered to earth receiving stations. This satellite circled the earth every 103 minutes (14 times a day) at a height of 900 km. It took 18 days to cover the globe (that is, to repeat the same orbit). Later *Landsats* have a lower, faster orbit—705 km, taking 16 days to cover the globe. See Figs. 4.5, 4.6, Plates 21, 26, 30. ((a) Image processed by NRSC, Farnborough, UK. (b) Reproduced by courtesy of Nigel Press Associates Ltd.)

to look at the sales of maps in relation to their appearance on television. In practical terms, the advice to someone setting off on a hill walk to 'take a good map' (as well as to wear proper shoes, take adequate warm clothing, and tell someone where you are going) is less likely to provoke terrrified and perhaps resentful paralysis. The maps at the back of the magazines in the airline seat pockets are almost certainly less neglected than they were.

The biggest mapping operation of all deals with the continuously telemetered output of the satellite-carried sensors and cameras (Chapter 4), which map at many wavelengths, at many angles of view, at all positions of the illuminating sun, and with a wide variety of data-processing methods (Fig. 7.8, Plate 21). This is where there are the most dedicated attempts to sort and index mapped items, using the biggest and best of our (unsuitably constructed) digital computer systems (Appendix A). Some of the work is concerned with the behaviour of the seas and weather systems, with prospecting for minerals, and with agriculture; but most of it is for military intelligence. In the preparation of this book we chose to restrict ourselves to the public domain, rather than to include material received in confidence that would have to be submitted for censorship, with the risk that speculation independent of this information might be classified. Most of the relevant knowledge of image-processing is, nevertheless, accessible through the channels of biology and biotechnology. Big makers of computers want to sell big computers everywhere, and therefore succeed in transferring many of the military advances to civilian users. If, as is possible and even probable, the main military secret is that there are no important secrets of information-processing principle, then this book will not have suffered. But it is necessary to bear in mind that there may well be significant lacunae.

Invisible pictures; infra-red; radio maps

The different kinds of radiation just referred to in connection with satellite-sensing comprise a considerable range in the invisible part of the spectrum, which will grow in extent and in importance. There are, as Sir Graham Smith and others have observed, 49 octaves in the spectrum of accessible electromagnetic radiation, all of which can be used for the taking and display of pictures. The visible spectrum, which is all that we can see directly, covers only one octave, or 2 per cent, of this rich array (Fig. 7.9). There are now detectors directional enough for us to make pictures using the radiation in many of the octaves. The astronomers are leading the way most broadly, with the military doing most in the immediate vicinity of the optical spectrum. Some parts of the spectrum are not very interesting on the terrrestrial scale, because the radiation is very penetrating, and goes through the things that we want to see instead of being reflected by them.

Invisible pictures; infra-red; radio maps 123

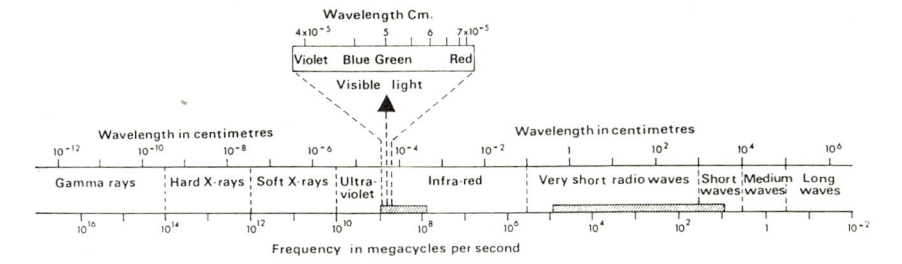

Fig. 7.9 The electromagnetic spectrum. (Reproduced from *Learning about space* with the permission of the Controller of Her Majesty's Stationary Office.)

The restricted range of wavelengths which we use need cause no surprise. Most of the sun's energy reaching the Earth is in the range of relatively short-wave visible, near-ultra-violet, and some infra-red radiation (Fig. 7.9). Over most of the spectrum our atmosphere surrounds us with a dense fog, acting as an impenetrable absorbent barrier to most other radiation, except the very short radio waves. Among the gases in the atmosphere, ozone absorbs the bulk of the ultra-violet, while carbon dioxide, oxygen, and water vapour absorb much of the infra-red. A good deal of light is also reflected back from the upper surfaces of the clouds. Accordingly, a wider range of radiation becomes useful to us only in satellites, which are above the atmosphere and can receive radiation direct from space, or when we are using artificial detectors to pick up man-made radiation such as broadcasting and X-rays, or the gamma-rays of radioactive minerals. None of this radiation played any part in the natural selection that determined what kinds of eyes we have.

The parts of the spectrum that have become important to us now are the infra-red and radio wavelengths, where suitable eyes would be very handy indeed. Infra-red enables us to discriminate between items that differ in temperature by no more than a tenth of a degree Celsius (Plate 22). In this way we can 'see' the level of liquid in an opaque tank by recording the differences in temperature of the outside surface of the container brought about by the differential heating and cooling by day and night of the filled and empty portions (Plate 23). Similarly we can trace heat leaks, and sometimes slight material leaks (Plate 22). Infra-red photographs taken from the air were used in the investigation of the disaster at the Chernobyl nuclear power station. Infra-red pictures enable us to look at many interesting things, and to 'see' in the dark. Living objects look quite different from dead ones (Plate 24). The middle range of infra-red radiation, reflected from the earth's surface, penetrates the atmosphere, and is intensively used by satellites (Plate 30).

Use of radio wavelengths makes it possible for us to 'see' far more stars and interesting objects in the sky, for there is a window in the fog at these wavelengths (Plate 25). The uses of radar and radio methods in detection and navigation are familiar to us all (Plate 26). There was one occasion in Scotland when a cooling tower collapsed at night. The first person to notice was the radar operator at the neighbouring airport, who rang the factory in a state of considerable bewilderment to ask what had happened. We can use sound pulses as well as light pulses (Plate 27, Fig. 8.7, p. 145). On the microscopic scale we can use not only photons, but electrons and nuclear magnetons (Figs. 6.6, 6.7, p. 89, 6.9, p. 90, 6.20c, p. 108). Our eyesight is improving, but we still cannot 'see' with gravitons.

We cannot add any more here about invisible pictures, but we shall certainly be seeing more of them as the special equipment becomes cheaper. 'Immortal, invisible, God only wise', maybe, but if it is only the 'light inaccessible' that hides Him from our eyes, He should be prepared for exposure. However, this will presumably be by non-electromagnetic light.

Graphs, bar charts, and pie diagrams

A now well-known figure once said, when he was Planning Officer of a large technological company, 'Never show the Board a graph. If it is simple enough to be understood by the accountants it will insult the engineers, and if it is interesting to the engineers it will puzzle the accountants. Keep it down to a comparison between twelve or so numbers, perhaps in a 3×4 table'.

Nowadays the use of graphs and charts on television has increased the general appreciation of the value of visual display. Most television news and economics programmes now include these. Many members of the audience may be interested spectators, who are affected by the oil prices or mortgage or foreign-exchange rates, but have little direct power to influence them beyond the ballot box. Others, as members of Boards or Committees, are shown graphs and charts when they are being asked for assent that calls for no action beyond agreement to continue (Fig. 7.2, p. 114). There are also operators on foreign exchange, commodity, and stock markets who work, not only with continually up-dated graphs which show trends, but with numbers the use of which calls for recollection of other trends. These latter can now be easily provided by the computer systems currently used for scrutiny of data and current prices. They have replaced the noisy telephone drama of the past. In fact, there is a considerable convergence between the operations of market brokers and control engineers, to whom we now turn.

Screen diagrams in control and guidance

We now come to the control panel, to be found in every car, television set, recording unit, aeroplane, factory of any size or complexity, or computer. This is one of the most interesting and important areas relevant to comparison between the utility of pictures and alphanumerics. Nearly everyone is in some way a control-panel operator, but there is much variety in the use made of the panel information put in front of us, and in the consequences of dedication or negligence. Drivers pay attention to the speedometer because of the law and the police, and to the fuel gauge so as not to get stranded. There used to be an oil-pressure gauge that gave information about engine functioning. An ammeter said how much electricity was being made or used. This helped with care of the battery, and with the avoidance of starting problems on cold mornings. However, so few people used the oil and electrical information that these were quietly dropped. Their places were taken by warning lights connected to cheap sensor systems. These simply notify a dangerously low oil-pressure, calling for immediate topping up of the sump, or a malfunction in the alternator/voltage-regulator requiring immediate repair.

Our control indicators (including our watches) usually started as dials and other analogue displays, because mechanical engineering preceded electronic engineering. One of the earliest lessons in information-handling is our childhood instruction in telling the time. The more recently arrived digital display may, in principle, be more precise; but its recognition and use employ a different brain function, and they cut out some useful approximate and anticipatory indications. The dial can show more clearly when you are *nearly* exceeding the speed limit, or 'ought to be getting ready to go' (Fig. 7.10). There is no absolute reason why numbers should be worse, but for many people they just are. Conditioning may be a factor, and the next generation may be less dial-oriented than we are.

There is a problem here, in that our present instruments are usually 'single-issue' indicators, leaving the question of priorities to the operator in a rather intellectual way. He may even have to walk about in order to scan all indicators that might call for action (Fig. 7.11). Many control rooms contain banks of cathode-ray-tubes ('television' screens). Now, however, it is possible to have a formalized picture on a single television screen summarizing the system being controlled (Fig. 7.12). Coloured indications (red for danger, green for safety, amber for caution) show the state of different control points. Thus he is able to take an instant 'top-down view' of the whole state of play, after which he can zoom in on the trouble spots, in order of seriousness, by calling up pictures of subsystems. (The use of such an improved control would probably have averted the Three Mile Island incident.) His master-picture may

126   *Pictures and words in everyday life and study*

———————————————— Control display ————————————————

Fig. 7.10 Single-issue analogue display. Pressure-gauge of steam train, showing clearly the desired limit. (Reproduced by courtesy of the BBC Open University Production Centre.)

Fig. 7.11 Control room where operators must walk about in order to scan indicators.

Fig. 7.12 Multi-issue display control panel. (Reproduced by courtesy of Unilever UK Central Resources Ltd.)

give him a dynamic image of the liquid flowing from one vessel to another, or of the level of the cooling fluid or lubricant, or of the positions of control rods. It can be a movie provided by computer graphics or a television camera. The television screen can also show the information as numbers or as dial readings. A soothing or raucous voice, or the familiar danger warnings of sirens and bells can be added.

It is a reasonable guess that most control information will in future be based on system diagrams or maps, with critical numbers clearly shown at the right points. Alongside the benefits of multiple-issue display, there is also the possibility of changing the control operations themselves, by introducing light-pens and touch-screens instead of switches, pedals, and levers. Much of the control response is passing to automated systems based on comparison with a simulated optimum and a machine knowledge of the operating rules, or even the underlying science. Where, in the end, does the human brain fit in?

Pictures in management, video-conferences, facsimile, time-lapse photography

Management can be regarded as an extension of control engineering. It introduces, as a central concern, the need to understand and persuade people. On the one hand, it requires the presentation of the state of play to one's colleagues, perhaps because one wants a larger share of the resources to invest, or because of new thoughts about future procedures and their development. On the other, it also requires understanding and good communications with customers, unions, and government about present products, efficiency, and operability, as well as about the future. So managers need a complex mixture of words, numbers, and pictures for their work. It is a little surprising that they still cling to numbers and words more than might have been expected. They could benefit in many ways from the informed and enthusiastic use of pictures, but this development has not got very far. Advertising contractors show some of the possibilities in the presentation of products.

So there is, and needs to be, much value analysis about the pictures that can best help when conversation is at a distance and requires the telephone. The facsimile machine, developed by the newspapers for sending photographs, is now cheap and can send reasonably complex pictures in a minute or so (Fig. 7.1, p. 114). The facsimile machine and the electronic scribble-pad (Fig. 7.3, p. 114) need only one telephone channel. Video-conferences are being used more (Fig. 7.2, p. 114), despite their cost, which will fall quickly when optical-fibre cables are available. The presentations can be on video-tape, which is an invaluable management information tool. It is one of the ways in which large numbers of people can be told something without assembling them into the subgroups needed for the showing of a film. The computer can make the transfer of pictorial information a two-way affair.

Another, more technical, management tool is time-lapse photography, which is so useful in biological perception. It can reveal movements in machines as well as in social operations over a long time-span, so throwing up patterns and oddities that are difficult to detect directly. The recognition of these can often solve problems. The technique has the added advantage of being picture-based, so that interpretation is as easy for the less articulate operator as for the articulate specialist. At the other extreme, high-speed photography is used to look at fast-moving machinery, so that events which happen too quickly can be slowed down and analysed in slow motion.

Pictures in the delineation and indexing of inventions; patent search

We now return to a topic touched upon briefly in Chapter 4. Four hundred years ago, in order to encourage the publication of information that would contribute to further technological progress, arrangements were made to grant a temporary monopoly in the use of an invention as a reward for the publication of its nature and means of achievement. This monopoly made it possible for the inventor to charge higher prices than would have been realized if imitators had been allowed to copy. In this way, too, he was enabled to recover his development costs. The conditions were that the information, published for others to build on, should be novel and not obvious. The invention, therefore, had to be precisely defined.

The problem is that alphabetical language is becoming increasingly inadequate as the principal—almost the sole—method of description and boundary-drawing. Patent specifications already contain information in non-alphabetical language, including diagrams and chemical formulae. These are included because in some situations they serve better than words to describe products, processes, plant, and equipment. Indeed, a complex machine is nearly impossible to describe alphabetically without a diagram as a back-up.

The range of inadequacy of alphabetical language is increasing. Electronic circuitry is the latest obvious case, the more so as it is now becoming three-dimensional. Sometimes, ideographic information is suitable for quasi-alphabetical procedure. Chemical formulae are as good as, and perhaps even better than, words for delineating non-engineered products, such as pharmaceuticals, agrochemicals, plastics, fibres, etc., and the process for making them. They can be translated into alphabetical form by methods such as that of Wiswesser discussed in Chapter 4. They can then be handled by ordinary storage and retrieval methodology. In principle, computer-aided design procedure can similarly handle a disciplined mechanical drawing; but the diversity of methods and protocols is already giving serious trouble. Moreover, the quantity of data in the computer description is several orders of magnitude greater than in a Wiswesser formula. If the drawing is impressionistic rather than engineered, then logical delineation, and the drawing of a boundary round what is claimed, become virtually impossible.

What, then, are the boundaries of the legitimate patent claims of an electronic firm for its latest random-access memory, an aero-engine firm for its latest core turbine or front-end fan, or a biotechnology company for its latest fermenter or micro-organism? And how is the ideographic description (in all three cases potentially crystal-clear to the informed scrutineer) to be indexed, filed, and retrieved in subsequent searches for prior art or for useful technical guidance? Accordingly, the increasing need for pictures, diagrams, photographs, and even video-tapes piles Pelion on Ossa in the new task facing the would-be improver of the patent and information system. The continued health of the patent system is likely to depend on an attack on the problem of delineation much more sophisticated than anything hitherto. Computers can help, but only when quite new patterns of thinking have been devised.

Unfortunately, no one is very interested, and the students of artificial intelligence, after a long period of quarantine from more orthodox scientists, are often thankfully turning to the practical and 'reputable' matter of expert systems, and away from complex linguistics. One starting point is the rationalization of computer-aided design through endeavours such as the Interactive Graphics Exchange System (IGES) of the US National Bureau of Standards. This could, in due course, create consensus around concepts of boundaries in machine designs. Another point of departure might be to describe processes as games, which may be more readily classifiable. Certainly, topologists and other pure mathematicians shoud receive encouragement. Small children in primary schools could generate useful ideas: the problem facing most adults is that they have great difficulty in discarding any of their existing intellectual furniture.

Other pictorial symbols and clichés

We have already touched upon the pictorial symbolism of chemical structural formulae, engineering drawings, electrical and electronic circuit diagrams, and various kinds of systems charts. In the street and the store, advertisers, who have been pioneers in the introduction of commercial art, use pictures alongside simple slogans about what is *Good For You*, what is *Liquid Engineering*, or what is worth *A Guinea a Box*. (Elaborate words and sentences

130   *Pictures and words in everyday life and study*

are out of place in this world.) Some commercial pictures are used for product description and technical selling. One Japanese robot-maker now puts out most of his product brochures in the form of video-tapes to show the products in motion and in action.

The merits of books, newspapers, cinema seats, and video-rental are extolled through stylized pictures carrying clichés about sex, violence, or other items that titillate. The novels of Jane Austen have been sold in paperbacks with highly imaginative pictures on the cover of wronged ladies in torn garments. The most sophisticated of the pictorial clichés are the cartoons—ranging from the immortal and scantily clad Jane in the *Daily Mirror* of the 1930s, 40s, and 50s to David Low's *Colonel Blimp* and the TUC *Carthorse*. Politicians are turned into cliché symbols. Sometimes a cartoon begins to look like a pictogram. *Mickey Mouse, Donald Duck*, and *Tom and Jerry* are the animated pictographic successors of the stills of *Billy Bunter* and *Dan Dare*. It is usually thought that this is the area of the non-intellectual and educationally insubstantial. Much snobbery is in evidence. It is acceptable for the child of upwardly-mobile or

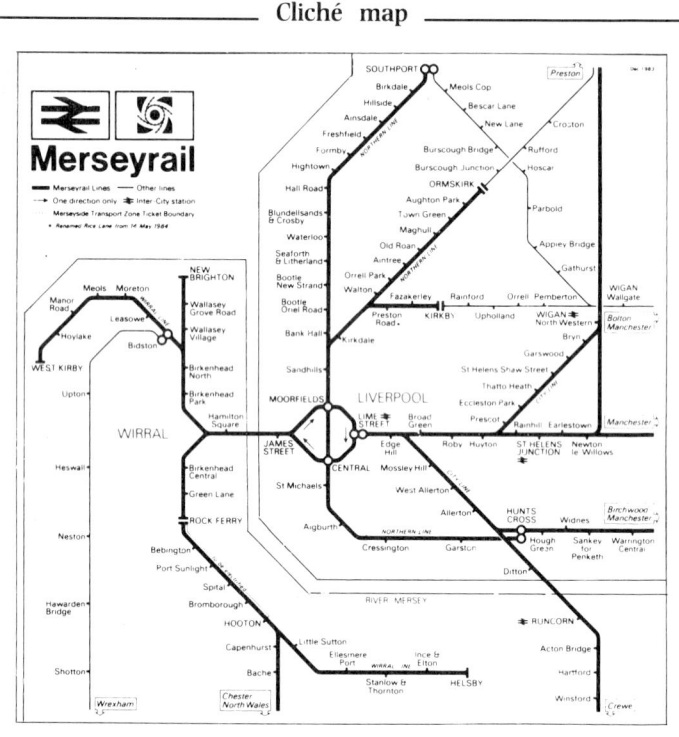

──────── Cliché map ────────

Fig. 7.13 Mersey Rail Map. (Reproduced by courtesy of the Merseyside Passenger Transport Executive.)

Other pictorial symbols and clichés

intellectual parents to like the *Winnie-the-Pooh* cliché, but not the *Woody Woodpecker*.

One useful pictogram is the simplified map of a railway line, road, or other everyday network, where the ordinary person's decision does not need precise geographical detail (Fig. 7.13). Such maps, started with organizations like the London Underground, the New York Subway, and various Metro networks, and are spreading rapidly. They may show clear instructions about evacuation routes in case of fire, or take the form of the 'you are here' maps of garden festivals or university campuses.

The range of pictures accessible to the ordinary person is impressive and growing, and there may be a spin-off or developments from any one of a number of specialist fields. The use of pictorial representations, symbols, and clichés can only increase.

8. The way ahead: the future of pictures

Methods and approaches

It is now necessary to consider what is likely to happen in the future, what is particularly desirable, what the obstacles are, and what should be done deliberately in the interests of priorities and major opportunities. Broadly, there are three possible approaches. The first is to concentrate on the technical factors, and to guess what incremental developments there may be in existing capabilities, and what new lines, products, and techniques may be invented. This approach may be inclined to throw emphasis on what will be feasible and may neglect, or pay insufficient attention to, social preference or the acceptance or rejection of innovations on the basis of their perceived utility. It is a matter of sad experience that the technocratic approach may overvalue cleverness and expertise, and underestimate the problems of customer preference and acceptance. The second approach is to list the places where we can expect an increased use of pictures, and the purposes for which they will be used, and to guess at plausibilities. This way emphasizes the realities of the home, the school, the hospital, the factory, and the service job, but may be liable to over-concentrate on the immediately possible, and neglect the longer term. It may also be a bit scrappy, in that very similar uses for pictures may have to be repeated several times, each for another environment and market.

The third approach, which we shall adopt, is to list the activities that pictorialization can encourage or create, and to examine likely new trends. This way of treating the issues can be easily understood, but may also need reinforcement if it is to pay sufficient attention to the forces of the market.

We can then subdivide under the following headings:-

(a) simulation;
(b) measurement control;
(c) demonstration, illustration, explanation, display of the invisible, decoration and entertainment, education;

(d) exploration and investigation, research, recording and storage of information; and

(e) communication and collaboration.

Training, guidance, and education through simulation

Simulation is not a new subject. Training has been done for a long time by putting people into a mock-up of an unfamiliar situation in which they are likely to find themselves. The resultant theatrical rehearsal is inevitably somewhat artificial, but it is enjoyable to many, generates enthusiasm, and undoubtedly works. Its oldest application is in the military field. Professional armies, and more particularly citizen militias, have always been given practice and confidence through TEWT's (tactical exercises without troops).

Mechanization of such exercises began with devices such as the Link Trainer of the Second World War for accelerating and cheapening the training of aeroplane pilots. It was an adjunct to, and not a substitute for, flying training. It also had merit in weeding out candidates whose slow reactions, lack of co-ordination, or perception defects made them unsuitable. As time has gone on, the realism of such equipment has increased, to the point where it can be used in situations where there is no possibility of flying training, as in the preparation for a satellite mission, where the next step after simulation is the actual rocket blast-off. In this training the moving pictures of the control panels, the capsule, and the total scene are indispensable. While simulation using numbers instead of pictures gives excellent results in the more leisurely and deliberate designing of manufacturing plant or processes, it is not appropriate for training people who must react instantly and in a sophisticated way to a seen image.

It is difficult to know in what proportion conservatism and expense have held back the transfer of training by simulation into the everyday civilian sector. There are excellent simulators for teaching merchant navy officers to navigate awkward and confined waters; but few people learn to drive a car in this way. It is interesting to consider where the boundary between simulation and reality should lie. Riding a bicycle is almost certainly a case where reality is cheaper. The same is certainly true for playing ball games such as lawn or table tennis, squash, badminton, cricket, or golf. Nevertheless, it might be that a simulator could be built to help train goalkeepers. After all, simulators for motor car driving and aeroplane flying have already become highly popular in the amusement arcades.

Potentially the most pervasive and versatile adaptation is now going ahead under the new name of 'interactive video'. This combines the sound and visual

Interactive video

Fig. 8.1 Training to interview by using video disc and computer. Frames from the disc.(Reproduced by courtesy of Interactive Information Systems Ltd.)

(a) Trainee identifies with, and role-plays, the interviewer.

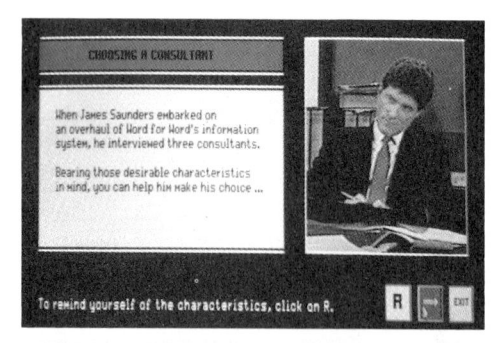

(b) Candidates are presented as options.

(c) The options are acted out on the film.

(d) Trainee evaluates candidates and makes the decision.

Training, guidance, and education through simulation 135

Interactive video

Fig. 8.2 A digital electronics course with off-screen work. (Reproduced by courtesy of Epic Industrial Communications Ltd.)

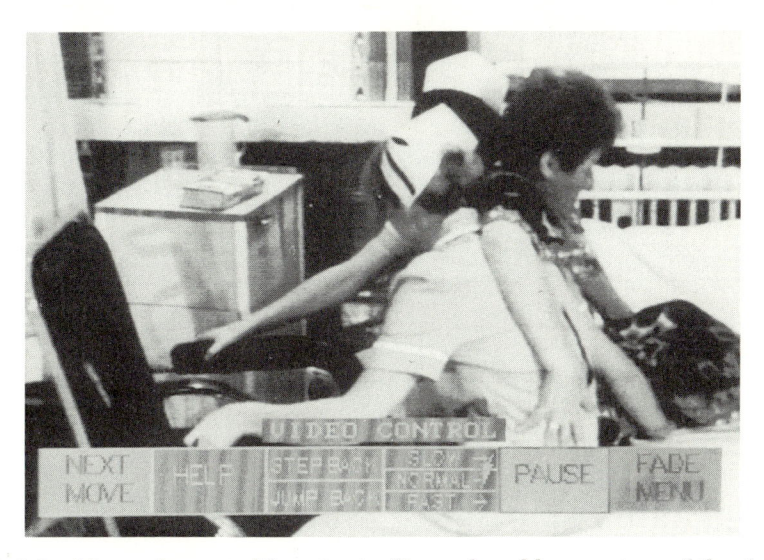

Fig. 8.3 Nurses learn to lift patients. (Reproduced by courtesy of the Audio Visual Dept, University College, London.)

information on a video-disc with the range of complex responses of the computer. It is a learning system that relies on dialogue between the student and the computer, avoiding the need for the presence of a teacher, but with the possibility of monitoring the student's progress. The system moves at a pace controlled by the student's responses, repeating or adding to information at the student's command, before moving on to the next sequence. It is likely to have considerable impact in education and training, and its use will bring a need for changes in teacher-training. Programmes make use of simple movies, animation, demonstration, dramatization, role-simulation, games, quizzes, and data retrieval. A simulation of, for example, an interview can easily be arranged (Fig. 8.1). The best programmes encourage much work away from the screen, to which the student returns from time to time for further instructions, for clarification, to input results, and to check conclusions (Fig. 8.2). Clearly interactive video can find uses in an immense range of situations (Fig. 8.3).

A special area which is being increasingly exploited is 'familiarization'. There are many situations where it is useful for a person to begin a task already armed with an understanding of a piece of machinery or the geography of his work place. The mechanic called upon to repair the engine of an unfamiliar car can brush up his knowledge by spending a few minutes with an interactive video. There are many needs for people in large industrial concerns to gain plant familiarization.

A special class of familiarization programme is known as 'surrogate travel'. The viewer of a television screen can explore a building, moving forward along a corridor—controlling his progress by a hand-manipulated joystick—glancing into a room, looking to the left, up, or down, peering closely at an object, even having a quick peep at the floors immediately above or below. On an oil rig, the worker newly come on board can already be familiar with its complex geography, and therefore ready to start work without further instruction (Fig. 8.4). Someone hiring a car in a strange city can have a dress rehearsal of any journey through unfamiliar streets to a new destination. The

Fig. 8.4 By using a 'joy-stick' you can 'walk', 'drive', or 'fly' through a pre-recorded environment. Here, you can walk an oil rig while in the office. The film (stills taken at 1-metre intervals) takes you all over the rig. One screen displays what you would see as you walk about, and the other shows where you are on the plan. (Reproduced from Statoil's Gullfaks Production Platform by courtesy of Visual Data Systems Ltd.)

Fig. 8.5 Point of sale (POS) video. Choose your food mixer without walking round the shop. (Reproduced by courtesy of Videodem Interactive Systems Ltd.)

Training, guidance, and education through simulation 137

Surrogate travel

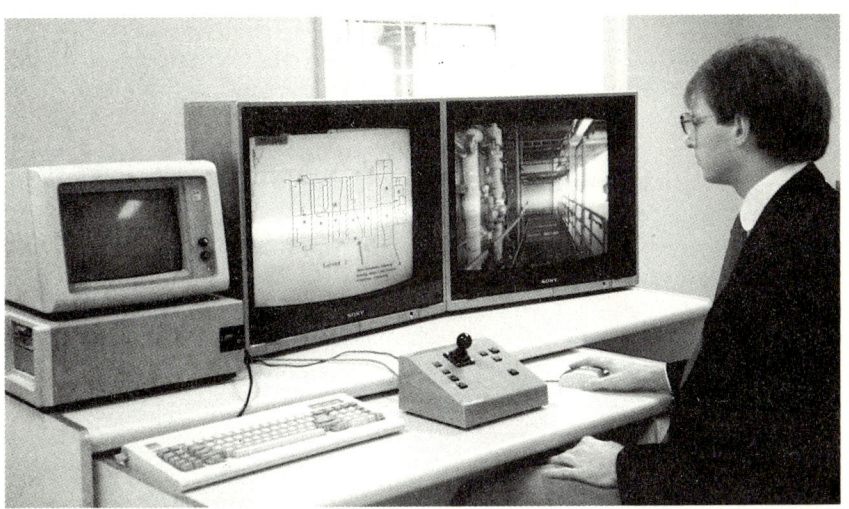

Fig. 8.4

Shopping by video catalogue

Fig. 8.5

time can be telescoped by using the 'fast forward' button (one of the most important inventions of recent times) on any easy section, involving, say, a motorway, with attention concentrated on the exit and entry ramps, the difficult junctions, and the places where traffic manoeuvring is complex. Landmarks can be seen, and the brain can perform its customary job of familiarization. The journey can be repeated as often as desired.

A specialized realm of surrogate travel is 'crisis management'. Security authorities, such as the police and the fire service to name but a few, need to know in advance of a crisis the detailed internal geography of an international conference centre, an office block, a factory, an underground station, or a nuclear power station. The technique was used by the Los Angeles Olympic Committee to prepare the security services for action in the event of terrorism.

The same method can be made available in the entrance to a gallery or museum. More generally, it can make use of the user's own video-player. A tape could be made of almost any activity, with explanation, important images, and alphanumeric guidance where helpful. For example, mail-order purchase could be moved away from the printed catalogue (which, let us be clear, many find entirely satisfactory) and be made more like proper shopping (Fig. 8.5). A video could provide guidance through a supermarket or shop. With an added telephone connection, a mannequin could be caused to walk about in any of the garments on offer, or a cook could prepare the dish provided in a food pack (As already noted, Japanese robot-makers demonstrate their products in this way.) Perhaps we should note that people go shopping partly for social reasons and to mitigate loneliness. The important point is that people are not entirely rational or logical, and a hierarchical algorithm that permits shopping by telephone may get the rejection that its name so richly deserves. They may prefer to do their video-shopping where they can see their friends and have coffee. The proposition 'Why not shop by telly where you can talk to your friends', though using the same technology, can be made more user-friendly, save a lot of exhausting walking about in the rain, and result in more value-conscious purchasing. Whether the actual buying is to be done in person, or by electronics, will also depend on technique.

Before going to a new opera, many would like a custom-commanded view of a critical video that contains explanation and commentary, gives access to important passages, and includes key libretto passages in words on the score. This could be done in several ways. For instance, booths could be set up in the foyer, where one could pay for any desired length of time, with prices highest in the hour before curtain up.

A catalogue of opportunities for interactive video would be tedious, and probably miss the areas that will become most important. The main point is that this use of pictures—with viewer participation—is not only certain to become important, but will be driven by forces that will usually guide it reasonably well, and finance the right kind of technology. A need for nonmarket activity may lie in the area of standardization, so that we all have to learn the tricks once only, and are spared the frustration of things that don't fit together. It may also be useful to think of uses in the fields of care and compassion, which will not easily find funds. Picture-systems for learning to be ten years old, or learning to be eighty, both offer food for thought.

Display of measurement and control

This subject follows from that of the last section, and deals with the question of the procedures to be adopted after the exploratory and diagnostic phase of simulation, when the survey of the road to be traversed or the goods available for purchase leads to action. How is one to drive or buy? As already mentioned in Chapter 7, pictures are increasingly used in the display of measurements of quantities relevant to driving, partly to cross language barriers by using pictograms in labelling, and partly to permit an instant overall survey of the position, in the interest of good priority-setting. Pictures of parts of the car, with matters needing attention in red, are helpful. A low fuel level, or oil-pressure, or battery charge-level, can thus be shown so as to nag gently. Speed can be given a different kind of display, because it changes so quickly and so much more often. If people were not all accustomed to steering-wheels and accelerator pedals, they might consider sitting behind a screen display and guiding the car by touching the screen at appropriate points in suitable ways. We suspect that the chorus of horror at such a suggestion is emotional rather than logical; but we also know that it is a waste of time to try to achieve such a change (if desirable) in less than a generation. People like to steer a dinghy with a tiller, a racing eight with two pieces of string, a ship with a wheel surrounded by handles, a power boat or a car with a steering wheel, and a light aeroplane with a foot-controlled rudder bar, because God started things that way and it is therefore right and proper. The fact that anything can now be steered by voice commands—'right a bit—no that's too much—straighten up a bit'—is of no interest to properly brought up children (Fig. 8.6). Pictures will advance in the driving of vehicles in response to curious forces, and there is no strong reason for interfering with preferences articulated through the market.

Shopping, too, will be done by methods responsive to psychology. There are some who would already be prepared to go to the coffee-video centre—sorry, center—and order goods by using a touch-pen to move a square saying '200 grams' over the picture of tea-bags made by a particular manufacturer (and optionally add the name, to make sure, and get a list of prices from all the stores in town, and pick the preferred one), touch 'ENTER' and move to the next item. At the end one would touch 'LIST' and 'DELIVERY CHARGE', get the response for those orders for each item, run the cashcard with the PIN number through the

A computer responds to voice commands

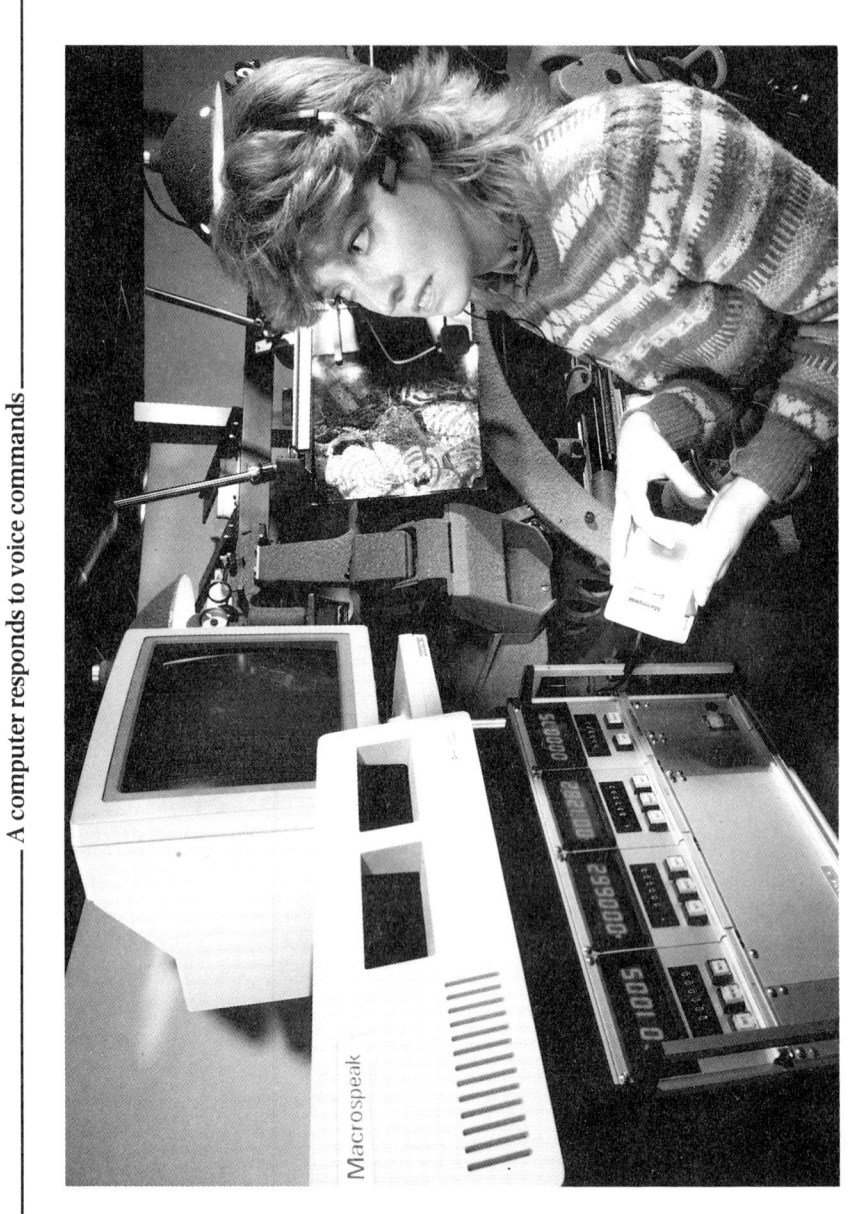

channel, and go home to await the deliveries or go on a collection round, or both. A large majority would find such a procedure inhuman and threatening, and perhaps do the round in person, paying as now at the checkout or the cashpoint. Some might do the geographically convenient shopping themselves, but order by post a major item available more cheaply from out-of-town.

A reasonable question at this point is 'What's so marvellous about pictures?'. Shopping, or enquiry about availability or prices, can already be done by telephone and mail order catalogue, and anyone who does this intelligently can save both effort and money. One reason for not doing so may be the inefficiency or unreliability of the responses, coupled with irritating and expensive telephone delays while enquiries are being made. Another option, which has not caught on, is the use of interactive view-data systems, such as Prestel. Would this have gone better with pictures instead of rather formidable lists and indexes?

We are inclined to suspect that pictures do constitute a quicker route for the acceptance of new command and control techniques as user-friendly; but they have to be good and well thought out. In principle, icons (pictograms) might be thought better than alphanumeric codes for computer commands. In fact, two things at present cause them to be rejected. The first is speed. The loading and use of the graphics programmes that are necessary for icons are usually slow for reasonably priced equipment. The second is obscurity. It is just as easy to remember that 'sc' or 'ws' or other brief code will command what you want as it is to work through several levels of little pictures which must also be learnt (Fig. 7.6, p. 118). Pictures of supermarket shelves might be a different story, particularly if access is through a sizeable and fast computer at the coffee-center.

Shopping and driving are good examples to think about in principle. Commercial forces might similarly create methods for choosing one's insurance, or banking, or travel arrangements. Pictures would be rather more symbolic, and words might often be preferred. But maps with times and prices on them could be quick aids for journey planning. Sheep farmers in New Zealand use a state computer system for planning and management of their flocks. Would such a system, for all kinds of agriculture, based on maps and pictures, be accepted in Europe? One could not round up or shear by touch-screen, but the fertilizer supply or contract-ploughing could be organized in

---

Fig. 8.6 The operator speaks into a microphone instead of using a keyboard. In this picture, the operator is entering mapping information. This system has been used for pilots to give commands to aircraft instead of operating switches and buttons. (Reproduced by kind permission of Clyde Surveys Ltd, Maidenhead, SL6 8BU, England.)

that way. Or are agricultural surpluses now an insuperable barrier to efficiency measures?

Other uses, however, may also be important, and less likely to be commercially driven. Would rational waste-management be assisted by arrangements for the custom-commanded collection of items capable of recycling or further use?

Decoration and entertainment, education, demonstration, illustration, explanation, display of the invisible

We can best begin this subject by comparing our actual uses of pictures with those well within reach at moderate cost. This shows clearly that it is conservatism, bewilderment, and custom that hold us back, and that, if technical advance is to have a noteworthy influence, it must be partly through the breaking of behavioural and psychological barriers. This was the secret of the advance of the movies between 1920 and 1940, and of television and video between 1950 and the present. The mere creation of capabilities that, on logical grounds, should meet a need will usually set in motion some slow change, taking several generations for completion. Our attention should therefore be strongly directed at people.

First, our murals have been unchanged by new capabilities. We hang paintings much as our ancestors did, and with no more innovation than that of the cheap reproduction. Some photographs are to be found, mostly black-and-white. (It is true that colour prints are not colour-fast for ever, but they are fast enough for use in a changing gallery.) There are endless possibilities for wall displays that can easily be changed electronically, for illuminated patterns, for fluorescent effects for special occasions, or for simulated scenes outside a window that looks out on a blank wall. If anyone doubts this, let him recall the new techniques that are being used by the more imaginative theatrical producers. Without abandoning the austere, which is being used as much as ever, they bring in fluorescence, solid items made from plastic foam, mobiles, new interaction between costumes and lighting, and means for simulating depth. There is virtually no trace of this in the home, where innovation is deployed on the house-decorating front in the improvement of paints (eliminating solvent bases, for example) and making decoration and cleaning easier. We do not even innovate much in lighting, although at discos and rock concerts we see patterns of coloured laser-light in the air that move to the music, and these may soon be used in the home (Plate 28).

Secondly, we continue to print pictures in books and put more, including coloured ones, in newspapers. But we do not send each other many photographs, even though we are all delighted when we get one or two. The

Decoration and entertainment, display of the invisible

family album and the scrapbook exist, but not on a grand scale. We send picture postcards as we have done for a hundred years; why only of scenery and buildings and not of our children? Just as we do not send audio-tapes, we have not yet begun to pick the best of the television we see to send to our friends abroad on video-tapes, even though it is easy. Is this due to bashfulness, nervousness, or a conviction that our correspondents would not watch them? To be sure, we have some problems with incompatible format, for example between Europe and the USA; but there are plenty of places where there is compatibility.

Thirdly, while the primary school projects that we have noted with approval in Chapter 3 are well illustrated with easily accessible material, doctoral dissertations could perhaps take more advantage of the scope for illustration. They are apt to contain graphs and austere diagrams, but generally few photographs. On the other hand, the needs for illustration vary considerably. A thesis on *Balzac as psychologist in Le Curé de Tours* would concentrate on French alphanumeric strings, and have scant need for pictures. On the other hand, one on *Patterns of coral growth in the Great Barrier Reef* is likely to consist of two volumes, one of text, the other of photographs, many taken underwater. University and school teachers vary greatly in the extent of their use of, and their need for, slides and film. Some sciences are heavily dependent on pictures as part of the means of research, as is clearly evident from no more than a reading of the popular science journals. A discussion on Keats's 'Ode to a Nightingale' would not be helped by photographs (though it might benefit from a sound recording). However, cell biology would be lost without the light and electron miscroscopes. Nevertheless, good transparencies or prints can put life into almost any lesson or talk, written or verbal. They can be bought cheaply at many art galleries, museums, and tourist centres. Cross-connections between subjects can be more readily grasped in visual form, as when a lecturer on ceramics can compare the textures of man-made porcelain with natural ceramic textures in rocks; or as in Plates 10, 18, Fig. 6.17 where we show the structure of DNA in a form understandable by the non-specialist reader.

Fourthly, instruction manuals vary between the dull but adequate and the utterly appalling. Nowhere is this more so than with computers, where nervous newcomers are picking their way into quite complex new capabilities, and are put off either by jargon or just by bad small mistakes that have to be researched. It is difficult to believe that the problem is not with one's own experience, but with the carelessness of someone paid to write and illustrate a text to sell hardware or software. Matters are improving a little; but we always need really good pictures of what the screen will look like after various commands, or how to put the printer ribbon in, or what kinds of chairs and postures are least tiring. Many of us would welcome a video-tape, and pay for it. Certainly, more people would do better in looking after their cars, cleaners, dish- or clothes-washers, or television sets and videos, if there were a video to help them. Why do cashpoints not have a video-loop running on a monitor just above them? To be sure, many customers have no difficulty; but some do, and more would use automated equipment if they were wooed a little. Post offices and banks already have videos that customers can watch while they wait, and which advertize available services. One of the needs is to recognize that there should be a variety of ways in. Some want it fast and some slow. The language that is appropriate for a trained engineer probably will not do for a poet or a peasant, and vice versa.

What new developments are to be expected? Will any of them lubricate the acceptance of pictures? Perhaps the first to note is the ability to see in places where light is not available. It is possible to produce pictures of good fidelity, from body-, head-, or organ-scans, with almost any kind of radiation (Chapter 6). Underwater sonar pictures of giant sand waves on the dark ocean floor are commonplace (Fig. 8.7), and night-time sensing with infra-red is becoming normal (Fig. 8.8). We now have many new kinds of 'spectacles' that enable us to see in new ways. We could soon be sold Japanese spectacles that can carry their own displays, different for the two eyes, and using radiation other than visible light: military night-glasses have been available for some time. This kind of display equipment ('spectacles') could also pick up radio or telephone messages; or one could sit down quietly and watch the three-dimensional television with no inconvenience to anyone else.

It seems unlikely that whole populations will be captured by these advances, on the model of television; but groups will be won. Security gear could interest householders and police, and it might be that football and tennis could be played in the dark by spectacled players, with infra-red television as the only means of watching, and with effort visible as higher body-temperature, or bad temper through a heated brow. The false-colour pictures of satellite astronomy (Chapter 5) may be the forerunner of some interesting and pervasive developments. It will be a bizarre novelty that will be the hardest driver. But again, what of those who do not want to be seen? Are they to be denied their invisibility?

The other limiting factors, apart from five centuries of habituation to words, are access and indexing. These are part of the subject of the next section.

Exploration and investigation, research, and recording and storage of information

There are three reasons why pictures may well acquire even more importance to reseachers, explorers, and investigators and therefore lend added urgency to the problems of sorting and retrieving images.

Exploration, investigation, research, recording and storage of information 145

—————————————— Seeing in the dark ——————————————

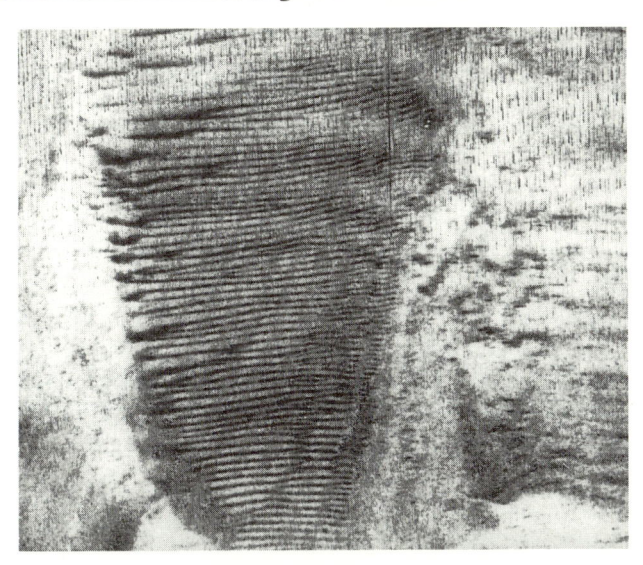

Fig. 8.7 Side-scan sonograph of a train of sand (or gravel) waves in the bottom of a slight trough, under the muddy waters of the Bristol Channel. About 18 m below sea-level, the waves are almost parallel, but interfinger and divide. Crest separation approx 10 m; height about 1 m. Area 596 m × 524 m. (Reproduced by courtesy of Arthur Stride.)

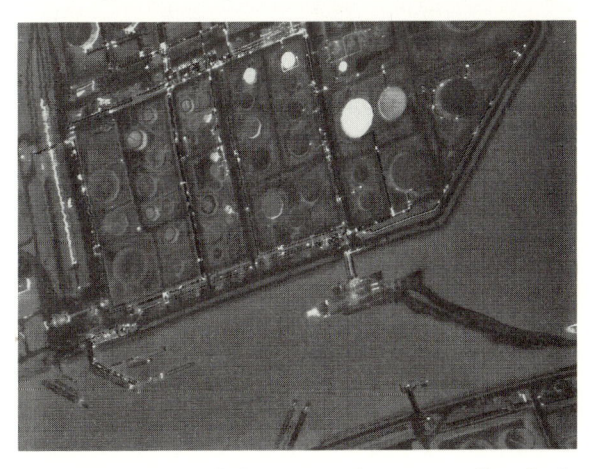

Fig. 8.8 Night-time infra-red linescan detecting illegal dumping from moored ship causing oil pollution, and heat loss from oil-storage tanks. (Reproduced by kind permission of Clyde Surveys Ltd., Maidenhead, SL6 8BU, England.)

First, the presentation of research results can gain from the use of computer and video images, still and moving, on the basis of arguments presented in earlier chapters. Secondly, the techniques in software for assembling sensor data of all kinds as pictures on screen, permit the instant viewing of experiments in many sciences. It is already true that the body-scanner has increased the doctors' devotion to pictures (Fig. 6.10), and that computer-generated molecular models are confirming chemists' commitment (Plates 10, 16, 17). This does not undermine the use of numerical or verbal analysis of observations, but adds to the power of quick 'look-and-see' techniques to help in the design of more quantitative experiments. 'Seeing in the dark' is especially important, as in the matter of inspection and the detection of faults (Plate 29). Examples here are such invisible things as the progress of creep in steel girders, the detection of cracks in the reinforced concrete of motorway bridges, and the monitoring of conditions in places where people cannot see or cannot go, as in the reactor core of a nuclear power station. Also the environmental and earth sciences are already deeply involved in the use of infra-red data collected from satellites (Plate 30).

As mentioned earlier, the third dimension is often helpful. It is important, for example, in the examination of geological structures seen in aerial photographs or in the construction of topographic contours on maps from aerial photographs. If two photographs are taken of the same object but from slightly different angles, as if with two eyes, they can be superimposed with the help of a binocular viewer or stereoscope (some people can do this simply by squinting) to give the illusion of three dimensions. Elsewhere, with the aid of a computer, outline drawings of serial sections through objects can be translated into stereoscopic pairs to give a three-dimensional picture (Fig. 8.9). In general, if the information from the two viewpoints is superimposed as a single picture, but differentiated in terms of colour or direction of polarization, then again the third dimension can be made available crudely, through the use of spectacles that pass different colours or directions of polarization through each lens.

Thirdly, in several sciences, the behaviour of 'average populations' has advanced far enough for it now to be necessary to give more attention to specially important members that have to be tracked down to their particular habitats. Examples abound in chemistry. The ions (charged atoms) in a crystal or in a solution, or the molecules in a gas or in an organic compound, are surrounded entirely on all sides by others of the same kinds and with the same strict geometrical or statistical distribution. Their chemical and physical properties can, therefore, be described in terms of their average values or bulk properties, or more strictly, by the equations of thermodynamics or statistical mechanics, which apply equally to them all. Such bulk properties are reasonably easily determined. But what about the ions at the two-dimensional surface of the crystal, for example at the interface between crystal and solution

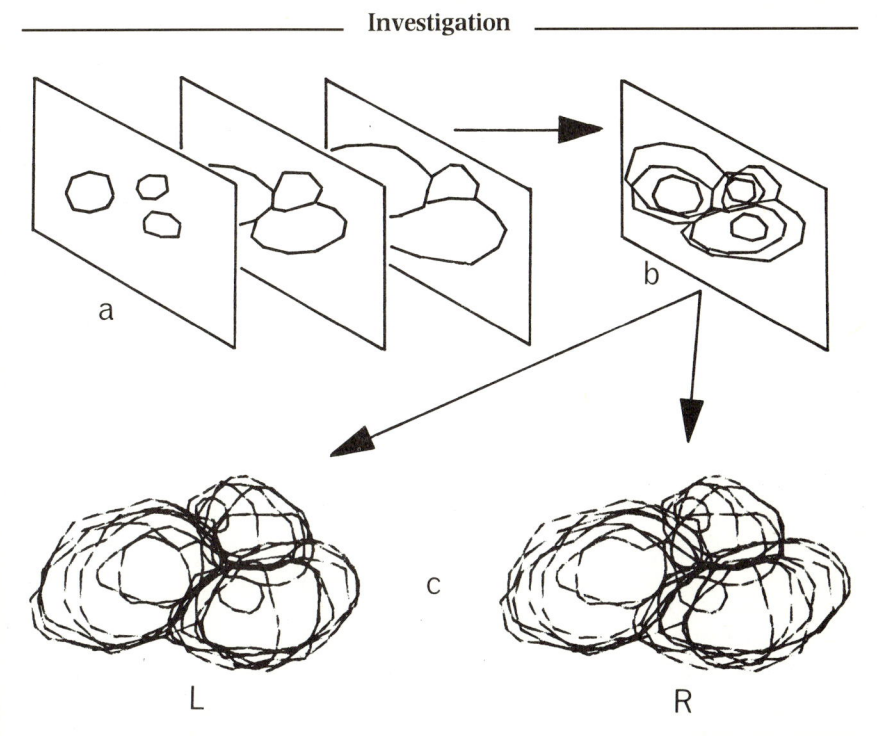

Exploration, investigation, research, recording and storage of information 147

─────────────────── **Investigation** ───────────────────

Fig. 8.9 Computerized stereoscopic pairs. Using an electronic pad adapted to a microscope, the overlapping outlines of three crystals are traced in serial sections of a rock (a). The result is changed into two images seen from slightly different directions (as from two eyes). Viewed through a stereoscope, the two tangles of black lines resolve themselves into the forms of three-crystals in the round. (Reproduced by courtesy of Robert H. Hunter.)

or gas? These are adjacent on one side to the geometrically arranged ions of the crystal, but on the other to a mixture of ions with molecules of water (dipoles), or to the molecules of gas. At this interface the chemical environment differs from that inside the crystal or in the solution or the gas. The properties of these interface ions and molecules are not bulk properties: behaviour is abnormal. Yet it is just at this interface that chemical reaction takes place. This special environment, though very thin (only a few ions or molecules thick) is immensely important. We urgently need to understand the properties of these specially reactive ions and molecules. Since, however, their properties cannot be described in readily determinable bulk terms (that is, in alphanumeric terms), the next best thing is to portray them in pictures.

Pictures have helped our understanding of a group of important chemicals known as catalysts. A catalyst is a substance, often present in relatively small amounts, which accelerates the rates of chemical reactions between other substances, but which itself remains unchanged at the end. They play a critical role in many chemical reactions, including large-scale industrial processes. As an example, uncatalysed chemical reactions between two gases mixed together depend on the number of molecules that exceed a critical activation energy and collide in the right orientation (bulk properties): all molecules have an equal chance of attaining reactive status. On the other hand, at the surface of a crystal in contact with a gas, single gas molecules may be adsorbed on to the lattice of the crystal, so that they are in a different environment from their free companions. In this situation some gas molecules may be much more able to react in special ways: the surface of the crystal is acting as a catalyst, so, again, pictures rather than the mathematics of averages are needed for explaining this special reactivity and improving its selectivity. There are many other cases where the mapping of experimental systems in this kind of way is increasingly important.

However, despite the breadth and importance of the research impact of the points just raised, the most significant area of all is that of new memory systems (nearer to that of the brain) to underpin revolutionary changes in image processing, sorting, and indexing. It would be good to discuss the kind of lines along which this work can logically be pressed, but this is very difficult. We are, quite simply, waiting for the breakthrough which will itself tell us what to do next. Until this comes we shall continue to flounder with unsuitable methods and equipment. It is true that we can improve present image-processing, and that much has expensively been done by the use of powerful and costly hardware like the Cray, which embodies the digital methods and the software that we have built up already.

The most practical way to help progress is to allocate money and thought to elementary approaches to practical problems where modest results will be useful. This at least will get people thinking about the issue. It will be necessary to ensure that they do not become imprisoned in incremental procedure, and thus implacably hostile to the unconventional. This suggests having some 'guessing money', paying for clever and unusual people, deployed within the main funding. The conventional work would be concerned with economical ways of transferring as many as possible of the military and satellite methods of image-processing into civilian use, in the indexing of television, film, and video. This could be done along lines useful to firms, with some of the funding coming from them so as to ensure their continuing interest. The pattern is that of Japanese 'pre-competitive research', but with a component that employs idiosyncratic people on adventurous science, pure in the first instance. Running such a team, and maintaining amity between the team and the individual operators, is by no means easy.

Communication and collaboration

Again, the problems are mainly human rather than technical. Preferred patterns of communication are highly individual. Some are good thinkers and expositors of their results through general and academic textual and numerical analysis and presentation. They may or may not be clear lecturers or participants in a discussion. Some are shy, and prefer to think things out alone, taking no part in active dissemination of information, and leaving the world to read their results or not as it chooses. Some prefer to write prose, some poetry. Some like to take photographs, and some to paint pictures or compose music. Few of the great creative artists are fully appreciated by their own contemporaries, although some attain some acceptance through chance (as in the case of Wagner's patronage by several people, including King Ludwig of Bavaria). Others, like Mozart and Dickens, become known by their performances of their own works and those of others. Performing artists, until recently, had only had their own generation as audience, and have been dependent on immediate impact. The time taken for the appreciation of scientific advance is variable. Because of some early mistakes Nobel Prizes are now mostly awarded two to three decades after the event. The potential of major technical advance normally takes a similar period to be fully recognized and significantly exploited.

Presentation and communication can therefore play a big part in the spread of new ideas, devices, and social practice. Sometimes it is called advertising, and employs highly-paid professionals, whose effectiveness can be, and is, measured: the results determine their prosperity. Sometimes it is called publication: effectiveness is more narrowly measured by peer groups who decide who shall be Professor, or a Member of a prestigious academy. Orthodoxy is more generally liked than radical innovation. Sometimes it is called, or is thought of as, entertainment: books, lectures, and television programmes bring in more money on this basis than because of their informativeness, although entertainment and information are by no means mutually exclusive. Sometimes it is called politics: through this channel emerged such innovations as democracy, factory welfare, general taxation, and social security.

The acceptance of new communication methods and channels thus depends primarily on how the customer or receiver is prepared to take his information, and what makes him act so as to implement it. Secondly, it depends on how many people, and which of them, are able to learn the new transmission methods. As long as some people can do this, the process will go ahead, although if the traditional presenters cannot manage the new methods they will suffer rejection. Consequently, the traditionalists will inevitably try to corrupt and disable the new means in all possible ways; but if the new is accepted, this rearguard action will do no more than delay change.

Optical-fibre transmission will soon provide channels that permit pictorially-based presentation and discussion to be as generally available for individuals or small groups as television is to everyone at large. The videoconference is still expensive, but will rapidly fall in price. Will this accelerate the use of pictures? Will it significantly replace face-to-face encounter? Must we wait a considerable time before we know how generally we want visual encounter at a distance? In the end, will there be something for everyone, or a technique for a minority? If it is for the few, will that select number be enriched and enabled, and others impoverished and to some extent disabled?

Telephones, letter-writing, and television have been with us for long enough for us to be able to guess the result of cheap pictorial conversation at a distance. We know that only a minority of people are good, keen, and effective letter-writers and telephone communicators. We note in the next paragraph the need to avoid the ill-mannered and intrusive way in which the telephone is so commonly used. We also know that politicians and others, who are good at speaking individually, or from a platform, often have great trouble with television, and need to learn the ropes (which some never manage). A former cinema actor seems to have been able to make considerable use of his training in the task of being head of a powerful state. So we can safely say that communication through pictures will require good techniques on the part of the cameramen and performers: the comunicators will have to recognize these skills and roles. Simply putting a static camera in the home or in a room, and then expecting to communicate well by behaving normally, will produce poor results.

This outcome is partly because the camera and the eye see differently, and partly because direct face-to-face communication uses many supporting factors. The telephone teaches us that preparation and enquiry as to the state of mind, preoccupation, and comfort of the listener are crucial. In his presence we can seek by many subtle means to know whether we have a chance of engaging his mind. From a remote position we are ignorant, and therefore likely to start off in the wrong way. Technique can help. Again, the use of visual aids needs skill. It is better to have none at all than to misuse equipment that is not understood. Lack of comprehension of the written or spoken word is a regular occurrence, and a request for clarification is usual and can be met by familiar procedure. No one is upset or irritated. There is no similar convention about diagrams or slides: it may seem embarrassing to admit bewilderment, and there may be impatience with enquiry, followed by frustration on both sides and a serious interruption of flow.

It is hard to know how many people will see the use of pictures as a new area for education and training, without which pictures may be more trouble than they are worth. The enlightened minority who take trouble may gain great advantages, which will be bitterly resented. Expect selective advance, and expect generally slow progress accompanied by Luddism.

Conclusions

The general upshot of this brief and necessarily superficial book is that we seem to be embarking on a new era of communication, comparable with the times that coincided with the invention of Arabic numerals and the phonetic alphabet about three thousand years ago. It took one millenium for the Indo-European alphabet to demonstrate its versatility through Greek, Hebrew, and Latin; another millenium for it to become the power base for a clerical church; and a third for it to permeate society. Two millenia passed before arithmetic was able to throw off the inconvenience of Roman numerals; after which mathematical progress and popular use came faster. So it may be that the human adaptation to the ready availability of our powerful range of pictorial techniques will be a thousand-year task. It may also be that this adaptation will be severely held back until we achieve quite new and unfamiliar memory methods and indexing. We have the circuses of television: how soon can these become relevant to learning, bread, manufacture, and services?

We have remarked in various places on the importance of the human factor in the acceptance of the new technologies and techniques, and we have asked whether some in society may be enriched while others, unable or unwilling to embrace the new developments, may be impoverished or even disabled. One may also ask to what extent will opposition to change reflect Snow's 'two cultures', and his portrayal of the arts and literary intellectuals as 'natural Luddites' in their reluctance to understand the scientific revolution? A woman physicist in McEwan's *The child in time*, frustrated by the indifference of the artistic world to the scientific, bursts out 'A scientific revolution, no, an intellectual revolution, an emotional, sensual explosion, a fabulous story just beginning to unfold for us, and you and your kind won't give it a serious minute of your time . . . Shakespeare would have grasped wave functions. Donne would have understood complementarity and relative time . . . They would have plundered this new science for their imagery. But you "arts" people, you're not only ignorant of these magnificent things, you're rather proud of knowing nothing.' Perhaps this attitude will be bypassed by the obvious practical advantages of the new devices, so neutralizing any tendency towards Luddism. Or again, will it be, in the light of what Strong suggested (Chapter 3), that the acceptance of the new ways of communication will reflect the division between the widely literate and those who can cope with little more than television and the tabloids? On the wider perspective, must we expect that the adoption of the new methods in the richer nations, which have the infrastructure to develop and service them, will further exaggerate the differences between the rich and poor in the world?

The diversity of the chapter subjects in this book carries a simple message:

With whom the future lies

Fig. 8.10 Primary school children in Devon communicating with school children in Tasmania by using electronic mail. For these children global, electronic, pictorial communication has become an everyday event. They are at one with today's élite commercial community. (Telefocus: a British Telecom photograph.)

Conclusions

there are many ways in which the adaptive and innovative processes can begin. Enough for us almost to be confident that at least one will provide the necessary opening. Throughout the book we have tried to indicate things to do in the near future, and to theorize and model mainly with this objective in mind. We may have been just a little too practical and pragmatic, and we may need to stand a long way back and theorize academically before the pragmatists get the right ideas to work on.

Scientists and technologists are already major and growing users of pictures, often dependent on them for further study; but they will have to stick to alphanumeric indexing of a laborious kind. Historians and linguists are likely to move into pictures as a relief from heavy concentration on words. Doctors and biologists will employ pictures even more centrally as a tool of the trade—and everyone will be simulating and controlling through images. But it is to the teachers that the main initiatives belong. They are the curators of the open minds, whose co-operation is needed if the best new balances between words, numbers, and pictures are to emerge. Perhaps it will pay us to pay them and esteem them better. Nothing could be worse for our purpose than a backward-looking, narrowly scholastic profession concerned with past specialisms, past methods of emphasizing the precisely-markable spelling and arithmetic tests, and too much love for the ancient cloister. We need the best of the verbal and numerical past, but it must be pliably united with the pictorial opportunities of the future (Fig. 8.10).

Appendix A: Recent work on computer graphics
Professor A. S. Meadows

Much contemporary work on sorting and identifying images relies on human capabilities of analysis, but speeds the process up by providing computer assistance. For example, geographical information systems are concerned with the simultaneous handling on computer of maps and data. The maps can be projected on a computer screen, and the operator can move the image in any direction, or zoom in and out, so controlling the level of detail available. Operators can be helped in their identification of objects by the way the computer presents the data. Colour coding is one popular way of showing data distributions on maps, though it is not always the best option. This will depend on the data to be displayed, which can vary from agricultural use of land to the distribution of water and sewage pipes.

An increasing use is now being made of graphical searching devices on maps. For example, the conventional symbol for (say) a church can be called up on the screen, and the computer asked to search its maps for any occurrence of this symbol. This works satisfactorily because the symbol looks the same wherever it appears. However, maps are now often being used in parallel with satellite or air photographs of the same terrain. Computer identification of objects on such photographs is much more difficult. A particular type of object does not always look exactly the same. Churches may have the same symbol on maps, but they look different on the ground. The favoured approach here is to 'teach' the computer to recognize a type of object. The computer has fed to it a whole series of images of similar objects, and is told that they are related. It then has the job of trying to extract the common features present in these images. Finally, a set of new images is fed in to see whether the computer can now effectively identify the chosen type of object. Its identifications are checked by the human operator for consistency. Recognizing churches is difficult. Land use—forests, grain, root crops, etc.—is easier to identify, and photographic images are already being analysed by computer for this purpose.

Fingerprint-matching is another area of image-analysis with which computers can now cope. Systems currently in regular use can automatically scan enlarged images of fingerprints from the scene of a crime, and seek to match them with fingerprints held in police records. The computer finds as many matching features as possible on each image, and estimates the similarity of the different fingerprints on a points system. At the end of the matching exercise, the fingerprints from the records receiving the most points are put in ranking order. The computer then displays to the operator the 'mug shots' and life histories of the people who have these fingerprints. Human checks can again be carried out on the fingerprint identification, but the vital requirement is that potential suspects should not be missed during the search. Hence, the systems developed for the police have to have a particularly high level of retrieval reliability.

One way of making the matching and identification of images more reliable is to enhance computer facilities by some form of artificial intelligence. Human knowledge of how to recognize and distinguish between similar-looking objects can be captured and employed in expert systems. This approach is becoming important in, for example, medicine, where large numbers of images from body scans of patients have to be evaluated rapidly.

These varied attacks on how to sort and identify images use ordinary computers. But the traditional computer does not handle image-identification as human beings do. It processes only one piece of information at a time, though, of course, it processes each piece very quickly. Human brains, on the contrary, handle many different items of information simultaneously. Whereas computers push all the incoming information down one channel, human brains allow it to flow down many parallel channels. The difference is especially significant for image-analysis, where the traditional computer approach is much less efficient than the human. A major recent development has been the appearance of 'parallel computers' which mimic this capability of the human brain. Though still experimental, these computers are already playing a vital role in the study of computer graphics handling.

The difference in approach at the simplest level is reflected in how a computer 'sees' the image of a human face. A traditional computer would have to scan the whole face, storing information on each part of it in a long linear sequence. A parallel computer divides the face up into a large number of tiny squares, information from each of which is recorded simultaneously. In consequence, when the same face is presented again, a parallel computer can recognize it quickly.

A real human face always changes slightly between presentations: because the face is seen at a slightly different angle, or the hair-style has changed, for example. The computer can be trained to recognize such differences, and to estimate for itself how likely it is that the same person is being looked at. Indeed, some of these computers, when fed part of a picture they have already seen, can identify the fragment, and reconstruct an image of the original picture. Parallel computers can increasingly link knowledge of images with other information. For example, researchers in the USA have developed a machine that can learn how to speak English words from written text.

Modelling computer operations on the way the brain works has now reached the stage where computers can be used to begin testing psychological theories of how humans identify images. The hope is that further developments incorporating the resultant knowledge will lead to greatly superior sorting and handling of images by computer.

Appendix B: The role of computer graphics in the exploration of chaos
Robin and Diana Bathurst

As we noted in the Preface, Duncan Davies had anticipated amplifying the section on the use of pictures in physics by adding an Appendix. Although he left no details of his intentions, we assume that he would have had in mind at least some reference to the new pictorially-dependent science of chaos. This is a method of studying dynamic systems, those that change with time, and it has evolved explosively through the use of computer graphics. It already has important applications, for example, in the understanding of the motions of heavenly bodies, the behaviour of elementary particles in an accelerator, the fluctuations of animal and plant populations, the beat of the heart and its fibrillation, the spread of epidemics, the nature of turbulence and the behaviour of weather, and, in the economic sphere, the fluctuating prices of commodities. Seemingly random variation of quantitative data which had hitherto been ascribed to unwanted 'noise' or experimental error has revealed deterministic chaotic behaviour.

The reader can readily observe chaotic behaviour with the help of an ordinary dripping water tap. Some of the most important research into chaos was carried out using this phenomenon. First adjust the tap until the water drips regularly. This is steady state like the oscillation of a clock's pendulum. Now cautiously and slowly turn the tap to increase the flow. At a certain point, before the flow becomes an unresolvable torrent, it will start to fluctuate in an irregular fashion, now fast, now slow, and wholly unpredictably. It is now chaotic.

Chaotic behaviour is not, however, random despite its unpredictability. It is constrained within definite patterns and it is these patterns that are so readily analysed with the help of the computer graphics screen. Selected elements of the algebraic equations that define the chaotic behaviour of dynamic systems are plotted on graphs. The results are not only informative but are commonly quite extraordinarily beautiful. Thus the beauty so often attributed to mathematics is made visible.

It was around the end of the 19th century that Poincaré began to express the laws of motion in the physical world visually, making use of geometrical forms instead of algebraic equations. From this work arose the science of topology, the study of how properties remain unchanged when shapes are deformed by twisting, stretching or squeezing, something that can be done so easily on a computer graphics screen. The extraordinary growth of the pictorial approach and its penetration into so many branches of quantitative analysis had to wait, however, for the development of the computer, above all for the graphics programme.

The reader can also experience chaos with the help of a pocket calculator and a simple algebraic equation such as children learn in school. For example, the equation that describes the annual fluctuation in size (x) of an insect population is: $x_{next} = rx(1-x)$ where r is the annual growth rate and x_{next} is the population in a year's time. Not surprisingly, as the population grows, food becomes increasingly scarce. So growth slows and the population eventually starts to fall. Later the food supply recovers and the population begins once more to grow. This cycle is repeated: the system is said to oscillate. (More physics-minded readers will recognize a feedback loop as x_{next} is constrained by r). Nevertheless, the property of this simple equation which has proved so disturbing is revealed if the calculations of x_{next} are repeated (iterated), just for a mere thirty iterations, starting, say, with $x = 0.2$. It is clear that for growth rates (r) between 1 and 3 the population settles down eventually to a steady state slight oscillation. Yet for rates above 3 the oscillations become increasingly extreme and, above 3.57, the values of x_{next} are unpredictable and fluctuate wildly and seemingly illogically. The system has become chaotic. Chaos is actually built into the numbers in the original equation.

The unpredictability of chaotic behaviour despite its deterministic organization is best appreciated by watching what happens when the appropriate algebraic relations are plotted as a graph on the computer screen. The first spots to appear seem to be distributed randomly. As iteration proceeds, and they become more numerous, a form begins gradually to emerge as the spots concentrate in certain regions. As more are added the outline of the form becomes better defined. It is impossible to predict where the next spot will appear, or the next thousandth. Nevertheless the region within which spots are likely to arrive becomes more and more sharply circumscribed. Disorder is channelled into a pattern with common underlying theme. Through millions of iterations the computer can expose the fine structure of a seemingly disorderly stream of data.

The iterations for x_{next} showed that only tiny changes in the original numbers in such non-linear equations can give rise, as iteration continues, to enormously magnified fluctuations as the behaviour becomes chaotic. This property is clearly of great practical importance. It has been remarked that the disturbance of the air by a butterfly's wing in, say, Kentucky might affect major

Appendix B: The role of computer graphics

changes in the cyclonic system over northern Europe. Equations of this kind that have built-in feedback loops commonly reflect systems where a body is subjected to more than one force. The squeals made by the loudspeaker while the chairman reads the minutes are caused when the sound from the amplifier is fed back repeatedly into the microphone. In certain conditions it is progressively magnified in successive feedback loops and the result is a hideous (chaotic) shriek. The significance for the normally regular oscillation of the heart beat is plain.

Pictures then are central to the *exploration* in this new science much of which is carried out through a voyage of discovery on a graphics screen. Only later may the results be amenable to rigorous mathematical proof. The technical terms reflect the pictorial basis of the science: phase space, mapping, orbits, sinks, saddles, tangles, bifurcations, cascades.

It would be wrong to ignore in this context the fractal geometry of so much chaos because this is pictorially so strikingly evident. A system is fractal if the geometric forms are similar at all scales. The branching pattern in a river system is fractal in the sense that branching rivers are fed by branching streams which are fed by branching brooks which are fed by branching rills which are fed by branching trickles. The same is true of the geometry of the arterial system of the blood. The outline of a coast can be treated as a fractal system because each peninsula is composed of a number of smaller promontories which are themselves made up of smaller bluffs, which are composed of tinier bumps which in turn may be constructed of sand grains. Although in fractals the same general pattern is revealed at all scales, it is never precisely copied. The pattern of clouds is fractal, which is why an observer finds it impossible to gauge their size. The forms resemble each other whatever the scale. Writers on the subject of fractals are apt to quote Swift:

> So, naturalists observe, a flea
> Hath smaller fleas that on him prey;
> And these have smaller fleas to bite 'em,
> And so proceed *ad infinitum*.

As Gleick put it in his book *Chaos* (which is an easy introduction for non-specialists), 'a fractal is a way of seeing infinity'.

Bibliography

Ash, J. E., Chubb, P. A., Ward, S. E., Welford, S. M., and Willett, P. *Communication, storage and retrieval of chemical information.* Ellis Horwood, Chichester (1985).
Attenborough, D. *Life on Earth.* Collins, London (1979).
Barlex, D. and Carré, C. *Learning through sharing images.* (Cambridge Science Education Series). Cambridge University Press (1985).
Bodmer, F. (edited Hogben, L.) *The loom of language.* George Allen and Unwin, London (1944).
Dawkins, R. *The blind watchmaker.* Longman Scientific and Technical, London (1986).
Diringer, D. *Writing.* Thames and Hudson, London (1962).
Gelb, I. J. *A study of writing.* Phoenix Books, Chicago (1952).
Gleick, J. *Chaos.* Heinemann, London (1988).
Hofstadter, D. R. *Gödel, Escher, Bach: An eternal golden braid.* Penguin Books, Harmondsworth (1980).
Hogben, L. *Mathematics for the million.* Allen and Unwin, London (1951).
Hogben, L. *Mathematics in the making.* Macdonald, London (1960).
Holt, J. *How children learn.* Penguin Books, Harmondsworth (1970).
McEwan, I. *The child in time.* Jonathan Cape, London (1987).
Michie, D. and Johnston, R. *The creative computer.* Viking Penguin, New York (1984).
Pask, G. and Curran, S. *Microman: living and growing with computers.* Century Publishing, London (1982).
Pochin, E. *Nuclear radiation: risks and benefits.* Clarendon Press, Oxford (1983).

Index

Advertising 33, 36, 38, 85, 127, 149
alphabet 9, 22, 23, 34
   Indo-European 9, 11, 19, 21, 23, 25, 151
   manipulation 3, 54
   phonetic 2, 19, 21, 56, 151
   picture 57–60
alphabetical order 56, 57
alphanumeric 125, 138, 141, 147
   cataloguing 53, 64, 66, 77, 92, 99
   string 3, 4, 28, 30, 75, 143
amino acids 99, 103–5, 109
analogue 57, 63, 65, 126
animation 33, 39, 65, 85, 119, 136
Arabic numerals 19, 27, 56, 57, 151
array processor 33, 64, 74
artificial intelligence 107–11, 129, 155
artists, role 87–92
astronomy 82
audio-tape 143

balance, words/numbers and pictures 14
bar
  chart 33, 73, 124
  code 56
binary
  arithmetic 3, 61
  code 11, 56–8
  numbers 26, 28, 63
biochemistry, pictures 98–101
biology, pictures 74, 86–111
biomorphs 106, 107
biophysics, pictures 98–101
body language 5, 41, 50
books 9, 13, 18, 21, 33, 36, 39, 41, 51, 65, 74, 96, 149

brain
  cryptology 64, 77, 107
  human 53, 61, 63, 65, 76, 92, 100, 109, 125, 148
  mammalian 53
  other animals 96
  recognition by 63
  science 107
  storage in 61, 64
  versus computer 61–4

camouflage 2
carrier 57, 58, 63, 67, 106
cartoon 33, 130
catalyst 148
cathode-ray
  screen 33, 69
  tube 2, 32, 35, 125
chaos 28, 82, 157–9
chemical communication 14
chemistry, pictures 69–85
chess 67, 110
childhood, pictures 38–52
chromosome 101, 103–5
cineangiography 92
cinematography 32, 92
clerks 2, 3, 9, 19, 23, 43, 151
clichés 129–31
code, genetic 101–7
codification 11, 18, 19–22, 28–30, 47, 57, 58, 64, 116
collaboration 133, 149, 150
colour 84, 92, 113, 125, 142
  slide 36, 38
communication 1–12, 84, 133, 149–51
  chemical 14
  corrupt, *see* deception
  drawing 39, 70, 74

communication—(cont.)
  non-recorded 6
  pictorial 12, 73
  primitive 19

computer 3, 4, 32, 35, 52, 56, 66
  animation 33, 85
  cassette-player 51
  confidence 12
  corruption, see deception
  digital 9, 25, 28–30, 58, 65, 74, 105, 119, 122, 125
  display 63
  in engineering 70
  games 46
  graphics 30, 31, 36, 47, 65, 69, 73, 77, 84, 97, 106, 127, 154–9
  indexing 11, 25, 26, 77, 110
  maps 119–24, 127, 154
  parallel 64, 110
  pictures 2
  recognition by 62
  retrieval by 77
  sorting 53, 58, 75, 110, 111
  storage 61, 63, 77, 155
  in surgery 92–5
  syntax 57
  in training 136
  translation 61, 67
  versus the brain 61–4
  video 65
conference, video 36
conservatism 133, 142
control, display 125–7, 132, 139–42
corruption of communication, see deception
Cosmos 78
Coulter Counter 78
Cray computer 148
crisis management 138
cryptography, military 32
cryptology
  brain 64, 77, 107
  genetic 101–7

data bank, pictorial 92, 111
deception 2, 50, 95
decoration, pictures 132, 142–4
demonstration 132, 142–4

diagrams 3, 28, 32, 40, 65, 69, 100, 150
dictionary 25, 41, 50, 54, 56, 57
digital
  computer 3, 9, 25, 28–30, 57, 63, 65, 74, 105, 119, 122, 148
  display 125
  electronics 135
  form 57–9, 61
  television 63
display 113, 114, 127
  analogue 126
  computer 63
  in control 73, 139–42
  digital 125
  of the invisible 132, 142–4
  in measurement 139–42
  screen 139
  video 46
distance learning 36, 48, 50
distortion, see deception
DNA 13, 100, 101–6, 115, 143
double helix, see helix
drawing 39, 41, 47, 70
  cartoon 33
  early 14, 19
  engineering 30, 43, 66, 69, 70, 73
  patent 129

Edinburgh Galaxy Machine 78
education 1, 26, 41–51, 132–9, 142–4
electrocardiogram 96
electromagnetic radiation 122–4
electron
  microscope 35
  stream 58, 63
encoding 2, 13, 24
engineering 27, 36
  computer aided 30, 36, 70
  design 32
  drawing 43, 66, 69, 70, 71, 73
  pictures 36, 69–85
entertainment, pictures 51, 112, 132, 142–4, 149
enzyme 99, 105, 115
everyday life 112–31
expert system 109, 110, 155
explanation 132, 142–4
exploration 133, 144–8, 159
explosion of pictures 32–5

Index

eye 5

facsimile machine, *see* Fax
false-colour 82, 144
falsification 2, 5
familiarization 136, 138
Fax 36, 115, 127, 128
film 148
filmstrip 47
foreknowledge, surgical 90, 91
fractals 28, 159
future of pictures 132–53

Galaxy machine 78
genetic code 14, 34, 101–7
gesture 6, 30, 34
graphics 3, 116, 141
  animated 39, 85
  computer 30, 31, 36, 47, 65, 73, 77, 84, 96, 127, 154–9
  video 84
graphs 33, 124, 143
guidance 125–7, 133–9

Halley's Comet 82
hearing, sense of 4, 5, 6, 14, 28, 41, 67
helix, double 101, 103, 106, 115
history of numbers, pictures, and words 13–37
holography 35, 69, 115
homophone 23
human factor 15

Iconclass 66
iconography 12
identification, *see* recognition
ideogram 2, 3, 9, 12, 21, 23, 25, 47, 67, 75–7, 115–18
illustration 132, 142–4
image, *see also* picture
  processing 74, 122, 154–6
  standardization 9
  transformation 57
  visual 6, 114
indexing
  language 25, 26
  numbers 53–68, 144

pictures 11, 28, 30, 33, 53–68, 73, 74, 77, 85, 92, 148, 151
  words 53–68
Indo-European alphabet 9, 11, 19, 21, 23, 25, 151
information
  analogue 57
  conveyed 1, 41, 51, 149
  in DNA 105
  flux 106, 107
  recorded 1, 6, 7
  transfer 41, 107, 108
infra-red 144–6
  maps 60, 122–4
intelligence, artificial 107–11, 129, 155
interactive
  teaching 48, 65
  video 35, 49, 50, 133–6, 138
  view data 141
Interactive Graphic Exchange 129
invention, patenting 128, 129
investigation 133, 144–8
invisible pictures 122–4, 144–6

journals, scientific 84, 143

Krebs' cycle 99, 100

Landsat 121
language
  barriers 70, 74, 116, 139
  Indo-European 9, 11, 19, 21, 23, 25, 151
  written 9
laser 78, 95, 142
lexicography 25, 53, 56, 105
librarian, word sorting 53
Link Trainer 133
Linnean taxonomy 92
liquid crystal 69

magnetic resonance tomography 90, 95
management 127, 128
maps 119–24, 127, 130, 141, 154
mathematicians 43

Index

mathematics 27, 28, 33–5, 40, 44, 157
measurement 6
 display 139–42
medicine, pictures 86–111
memory
 brain 11
 computer 9, 25, 148, 151
 human 76, 148
 mammalian 76
 molecular 77, 107–11
microscope
 electron 35, 100, 105, 143
 illustration 92, 112
 light 34, 78, 100, 105, 143
military ideogram 116
model 82, 83, 146, 156
 formulation 53, 96, 100, 101
Moire fringe pattern 92
molecular
 beam epitaxy 82
 communication 5
 memory 11, 107–11
 senses 41
 sorting 101
 storage 101
 structure 67
molecule
 as carrier 58, 64
 giant 77, 98, 99, 101
 organic 101, 105, 146
music 5, 28

nerve network 100, 108
neuron 64, 109
newspapers 9, 33, 35, 36, 39, 43, 84, 112, 142
numbers 3, 4, 6, 8, 9, 11, 18
 balance with pictures 14
 in childhood 38–52
 history 13–37
 sorting and indexing 53–68, 74

optical fibre 150

parallel pathways 64
parallel processing 64, 109–11, 155
particle tracks 78, 79
patents 28, 54, 128, 129
pathology 95, 96

pattern recognition 57, 63, 107, 110, 128, 154–6
photography 2, 33, 35–7, 47, 69, 88–92, 112, 129, 143, 149
 black-and-white 32, 39, 51, 142
 colour 32, 38, 39, 142
 high speed 128
phonons 41, 58
photons 41, 58, 63, 124
physics 95–6
physics, pictures 69–85, 157
pictogram 1, 2, 3, 8, 21, 41, 75, 115–18, 130, 131, 141
pictures 1
 in astronomy 78–82
 balance with words and numbers 14
 in biochemistry and biophysics 98–107
 in biology 86–111
 carrying information 2
 in chemistry 69–85
 in childhood 38–52
 codification 11, 18, 19–22
 in collaboration 133, 149, 150
 in communication 133, 149, 150
 in control 132, 139–142
 in decoration 132
 in demonstration 132
 display 113, 114
 in display of the invisible 132
 in earth sciences 78–82
 in education 46–52, 132–9
 in engineering 69–85
 in entertainment 132
 in everyday life 112–31
 in exploration 133, 144–8
 explosion 32–5
 facsimile 127, 128
 in guidance 133–9
 history 13–37
 in illustration 132
 indexing 11, 28, 30, 33, 53–68, 74, 77, 85, 92, 148
 infra-red 122–4
 in investigation 144–8
 in management 127, 128
 in measurement 132, 139–42
 in medicine 86–111
 in patents 128, 129
 in pathology 95, 96
 in physic 95, 96
 in physical sciences 75–82

Index

in physics 69–85, 157
radio 122–4
recording 9, 18, 30, 36, 133
in recording 144–8
representational 13
in research 144–8
retrieval 11, 33, 77, 144–8
in simulation 133–9, 142
sorting 53–68, 74, 85, 92, 148, 155, 156
storage 77, 144–8
time-lapse 127, 128
in training 133–9
in video-conferences 127, 128
pie diagrams 33, 124
pixel 11, 12, 57
Prestel 12, 141
priests 21
printing 2, 9, 11, 18, 43
protein molecule 75, 76, 99, 100, 103, 104–6
psychiatrists 47
psychologists 47, 107

Quantel 113
QWERTY keyboard 11

radar 124
radiation, electromagnetic 122–4
radio 48, 51, 95
  maps 122–4
  signals 57
  waves 124
reading 2
recognition 57, 59, 61–3, 107, 110, 116, 154–56
recording
  of pictures 18, 30, 32, 36, 133, 144–8
  of words 23–6, 34
  video 95
remote sensing 60, 66
replication 13, 101
reproduction
  of information 101
  of species 14, 86
research, pictures 133, 144–8
retrieval 1, 11, 33, 59, 76, 77, 128, 129
RNA 103
road signs, *see* traffic

satellite 48, 57, 59, 60, 66, 78, 80, 95, 110–24, 133, 146, 148, 154
scanning 30, 35, 61, 91, 97, 126
  electron microscope 88, 89, 109
  in television 2, 11, 32, 57, 58
  tunnelling microscope 82
science journalism 84, 143
screen diagrams 125–7
senses 4–11
shopping by video 137–9
sight, sense of 4, 5, 6, 14, 28, 41, 67
simulation 132–9, 142
slides 47
smell, sense of 1, 4, 41, 67
sonar pictures 144, 145
sonic log 81
sorting
  numbers 53–68
  of organic molecules 101
  pictures 1, 53–68, 74, 75, 85, 92, 111, 148, 155, 156
  words 25, 53–68, 75
sound 4, 5, 6, 14, 28, 41
spectroscopy 75, 80, 82
standardization 9, 12, 18, 19, 23, 26, 34, 139
  absence of 59
stereoscopy 98, 146, 147
storage
  brain 61
  computer 61, 63, 77, 155
  in organic molecules 101
  patents 128–9
  pictures 1, 77, 133, 144–8
strings, linear 2, 11, 27
study 112–31
supermarket 112, 141
surgery, pictures 90–5
surrogate travel 136–8
symbol 11, 41, 129–31, 154
syntax, pictorial 57–60

tapes 3, 4, 27, 33, 35, 36, 51, 64–6, 74, 85, 96, 127, 129, 130, 143
taste, sense of 1, 4, 41
teachers, in school 38, 46, 153
telecommunication 73
telemetry 33
telephone 51, 52, 150
  switching 54–6

Teletext 12
television
  in advertising 85
  in astronomy 82
  in brain analysis 64
  camera 57
  in childhood 38
  closed circuit 95
  in communication 2–5, 35–7
  in display of maps, charts, graphs 119, 124, 125
  in everyday life 112, 113
  future development 142–4, 148–151
  in learning 12, 46, 47, 50
  in surrogate travel 136
  use by Open University 65
theory, formulation 53
three-dimensions 9, 30, 35, 65, 67, 74, 78, 99, 101, 115, 146
time-lapse photography 127, 128
tomography 90, 92, 105
topology 158
touch, sense of 1, 4, 41, 67
traffic signs 116–18
training 133–9
transfer of information 107, 108
translator routines 61, 67
tunnelling microscope 82

video 35–7, 48–50, 52, 142, 144, 146, 148
  camera 3, 11, 96
  catalogue 137
  computer 65
  conference 36, 115, 127, 128, 150
  disc 134, 136
  display 46
  graphics 84
  interactive 35, 49, 50, 133–6, 138
  library 66, 76
  loop 144
  phone 115
  recording 95
  rental 130
  reports 82–5
  shopping 137–9
  tape 3, 4, 12, 33, 35, 36, 51, 64–6, 74, 85, 96, 127, 129, 130, 143
voice command 139, 140

way ahead 132–53
weather forecasting 119, 120, 122
Wiswesser Line Notation 67, 75–77, 128, 129
words 1, 3, 4, 9, 11, 18
  balance with pictures 14
  in biology 86–111
  in chemistry 59–85
  in childhood 38–52
  in education 46–52
  in engineering 59–85
  in everyday life 112–31
  history 13–37
  in medicine 86–111
  in physics 59–85
  in study 112–31
  sorting and indexing 53–68
writing 2, 3, 4, 28, 43, 84, 150

xerography, genetic 101–7
X-ray 92, 95, 99, 101, 123

zero 19